D0205040

Quaternary Landscapes

QUATERNARY LANDSCAPES

Edited by
Linda C. K. Shane
and
Edward J. Cushing

In cooperation with the
Newton Horace Winchell School of Earth Sciences
and the
Institute of Technology,
University of Minnesota

University of Minnesota Press, Minneapolis

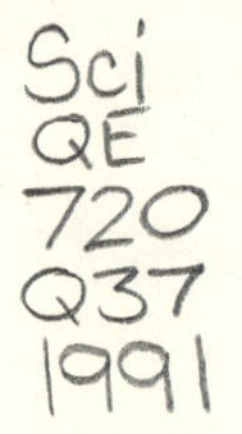

Published by the University of Minnesota Press
2037 University Avenue Southeast, Minneapolis, MN 55414.
Printed in the United States of America on acid-free paper

Cover illustration by Donald T. Luce.

Library of Congress Cataloging-in-Publication Data

Quaternary landscapes / edited by Linda C. K. Shane and Edward J. Cushing.np
p. cm.
Papers from a conference held at the University of Minnesota, sponsored by the National Science Foundation, Climate Dynamics Program and various departments of the University.
Includes bibliographical references.
Includes index.
ISBN 0-8166-1943-3 (alk. paper)
1. Paleoecology—Congresses. 2. Geology, Stratigraphic—Quaternary—Congresses. 3. Paleontology—Quaternary—Congresses.
I. Shane, Linda C. K. II. Cushing, Edward J., 1933- .
III. National Science Foundation (U.S.). Climate Dynamics Program.
IV. University of Minnesota.
QE720.Q37 1991
560'.45—dc20 90-49455
CIP

The University of Minnesota is an equal-opportunity educator and employer.

To Rex Wright
and
To Rhea Wright, whose cheerful, perceptive, and welcoming spirit and energy even in her last days were a constant joy and encouragement to all who knew her.

Contents

ACKNOWLEDGMENTS

We gratefully acknowledge the many people who helped make the symposium and this book possible. The planning committee involved people from three University of Minnesota departments: (1) Geology and Geophysics: Daniel Engstrom, Peter Hudleston, Olaf Pfannkuch, Linda C. K. Shane, and Joseph Shapiro; (2) Ecology and Behavioral Biology: Robert Bright, Edward Cushing, and Eville Gorham; and (3) the Center for Ancient Studies: Peter Wells. Also involved were Eric Grimm, now of the Illinois State Museum, and Robert Baker of the University of Wisconsin-River Falls. The symposium and this volume were sponsored by the National Science Foundation, Climate Dynamics Program, and the following ten units within the University of Minnesota: the departments of Botany, of Geology and Geophysics, of Geography, and of Ecology and Evolutionary Biology; the Center for Ancient Studies; the Institute of Technology; the Colleges of Biological Sciences and of Liberal Arts; Professional Development and Conference Services under Continuing Education and Extension; and the Office of Provost and Academic Affairs. The Department of Geology and Geophysics has provided valuable additional financial support for postage, phones, copying, and student technical work during the editing phase of production. The thematic landscape illustration is by Donald T. Luce, Associate Curator of the James Ford Bell Museum of Natural History. Very helpful outside reviews were given by Lou Maher and Jim King. Greg Liu, cartographer of the Geography Department, coordinated all the final work on the figures, much of which was done by Lois Eberhart and Philip Schwartzberg. Robert Grogan prepared a valuable index. In addition, many colleagues (inside and outside the LRC) have helped with the myriad of details this whole project has involved. These include Elizabeth Almgren, Ann Dickason, Sheri Fritz, Paul Glaser, Peter Hudleston, George Jacobson, Judith Jones, Sharon Locke, Howard Mooers, Orrin Shane III, Thomas Trow, William Watts, and Herbert Wright, Jr. Finally we thank Chip Wood, Lisa Freeman, and the wonderful production staff of the University of Minnesota Press, whose help and consideration were essential to this project.

Conventions

Several conventions concerning expressions of time are used in this book and are listed here. Dates are given in uncorrected radiocarbon ages unless specified otherwise. Note also that in Patty Jo Watson's chapter, dates are given with the Christian-calendar alternative in parentheses because research in archaeology uses those terms as or more frequently than the terms used by geologists.

ka = thousand(s) of years ago
yr B.P. = year(s) before present
b.c.e. = before the common era
a.c.e. = after the common era

INTRODUCTION

The late 19th and early 20th centuries were a time of rapid evolution in the natural sciences. Physics, chemistry, geology, and biology all blossomed; new subdisciplines sprouted and grew quickly. Specialization continues, but in the second half of this century interdisciplinary studies attract more and more scientists. Nowhere is this more true than in Quaternary research, the study of changes in climate, oceans, topography, biota, and human evolution over the past one million years or so. Understanding these changes provides keys to prediction of the future and a better understanding of ourselves as human beings.

The Quaternary period is the most recent geologically and has the most complete geologic record. On its varied landscapes humans evolved biologically and culturally from minor, passive players to dominant agents of change throughout the globe. The plants, animals, landforms, and climates are familiar to us. As our knowledge and technology have grown, our vision of our world has expanded also, until now we can both see it as a whole and begin to comprehend its workings. But global processes are slow, and our observations span too short a time to permit our full understanding. The records we need are archived in the deposits of the Quaternary. The information found there allows us to synthesize images of landscapes at various places and times in the past. As those images accumulate, we combine them to see dimly, as if viewing a badly flickering movie in an undarkened room, the rhythms and the evolution of our world and, with it, ourselves.

Research in the 18th and early 19th century was interdisciplinary almost by definition, because scientists could have universal interests and knowledge. Within this generalized framework, two fundamental observations were made. First, many geologic strata contain ancient plants and animals, often very different from biota observed today. Second, large parts of Europe and North America were covered with massive glaciers relatively recently, suggesting major changes in climate in the near past. The implications of these startling discoveries were gradually recognized. In Europe, for example, where the cutting of peat for fuel generated many sections within basins that had been created by the glaciers, clear stratigraphic changes were seen that suggested interpretation in terms of changing climate. This led to the study of plant fragments within the peat and ultimately to the realization by von Post of the scientific value of windborne pollen. The stratigraphic analysis of other microscopic fossils, such as diatoms, insect parts, shells, and ever smaller fragments of larger plants and animals, is now a routine part of interdisciplinary work in Quaternary studies.

The goal of this book is to demonstrate the broad range of research encompassed by Quaternary studies and the value of interdisciplinary approaches to understanding Quaternary problems. The six chapters were originally presented in May 1988 at an international, interdisciplinary symposium held to celebrate the role of the Limnological Research Center at the University of Minnesota as a leader in interdisciplinary Quaternary research and to honor Herbert E. Wright, Jr., at the time of his formal retirement. The Limnological Research Center (LRC) was founded in 1960 as a unit within what is now the N. H. Winchell School of Earth Sciences, Department of Geology and Geophysics, with funds from the Hill Family Foundation. In 1963 Wright became its first permanent director and Joseph Shapiro its associate director. Wright brought to the directorship expertise in Quaternary geology and international experience in successful collaboration with archaeologists and paleoecologists. From the beginning the research goals of the LRC were cooperative, international, and interdisciplinary, including studies of the contemporary biological and physical processes in lakes and of the history of lakes and their surroundings as preserved in their sediments.

The symposium honoring Wright was organized to illustrate the breadth of Quaternary research that evolved in the LRC under his leadership. That breadth is illustrated graphically by the landscape pictured on the front cover of this volume. Parts of that landscape, like pieces of a jigsaw puzzle, are linked to each chapter. The central concern of the symposium was to represent the diversity of fields that must be understood to investigate Quaternary landscapes. The topics were chosen to represent disciplines to which the LRC and Wright have made major contributions. The contributors were invited because of their international reputations as researchers and speakers. Each was asked to present a paper reflecting on the state of the art in his or her own field—where it has been and where it is going. Like the pieces of the landscape puzzle, the chapters merge to provide views of how the ever-changing Quaternary landscape can be recreated.

The opening chapter by Patty Jo Watson represents the field of anthropology and archaeology. In the early 1950s, she participated with Wright in Robert Braidwood's project on the origin of agriculture in the Near East. She has since worked extensively in North American archaeology, encouraging full recovery of floral and faunal remains as well as the traditional cultural artifacts. In both regions a simple question—"how and why did the domestication of plants and animals begin?"—has itself become an interdisciplinary investigation. Not only are models of changing physical landscapes relevant, but theories of human social and cultural change are critical. Here the natural and social sciences blend seamlessly.

In the second chapter Jan Mangerud represents the field of glacial geology and geomorphology, Wright's original area of research. Mangerud summarizes recent developments in the Pleistocene history of northern Europe, especially apt because of Wright's long dedication to bringing awareness of that complex history to Quaternary geologists in North America. In his own interdisciplinary synthesis, Mangerud fits together evidence from lithostratigraphy, pollen stratigraphy, and oxygen-isotope stratigraphy, from both terrestrial and marine deposits, to build a consistent picture of the changes northern Europe has seen in the past 120,000 years.

In chapter 3, Richard Clymo discusses peat and peatlands, another field to which Wright and colleagues in the LRC have contributed much. The growth of peat and the development of extensive peatlands often result in striking patterned landscapes. To understand these enigmatic patterns requires detailed knowledge of the plants that grow on the peatland, the microbiology of degradation of the peat that accumulates as the plants die, the physical and chemical properties of the peat throughout its thickness, the geologic setting of the peatland, and the effects of climate and groundwater flow on the hydrology of the peatland system. Clymo develops a model of peat accumulation that includes these factors. The quantitative aspects of his model point clearly to the kinds of observations that must be made to increase our understanding of peatlands.

The fourth chapter, by James Ritchie, represents the field of paleoecology. Ritchie has always stressed the relationships among plant communities, the pollen and other fossils they produce, the climate, and the physical characteristics of the landscape, and here he illustrates them with an example from the Sahara. He further considers some potential contributions of plant paleoecology to community ecology and biogeography, with promising examples from recent work.

Chapter 5, by Richard Battarbee, represents the field of paleolimnology. Wright has long encouraged the use of lake sediments as a limnological archive. During a year's sabbatical at the LRC, Battarbee stimulated research into the use of saline lake sediments for paleoclimatic reconstruction. Much of his research involves the application of paleolimnological techniques, particularly diatom analysis, to such current ecological problems as lake acidification. He reviews the methods and problems of this type of analysis and underscores the value of long-term records in understanding human impact on watersheds and lake basins.

The sixth chapter is a joint effort by John Kutzbach and Thompson Webb III, representing the field of paleoclimatology. Both were initiators, with Wright and others, of COHMAP (Cooperative Holocene Mapping Project). The chapter summarizes the history of Quaternary paleoclimatology, which stimulated this global, interdisciplinary research project, and some of the results of the project, which has attempted to test climatological models with data from Quaternary sediments. Kutzbach provides lucid explanations of some of the climatological processes that can be recognized in Quaternary deposits, and Webb describes the implications of the extensive set of data on pollen stratigraphy that he compiled from contributions by many people. Their chapter illuminates the need to match the spatial and temporal extent and resolution of data with the scale of the processes that are being modeled.

The COHMAP project was a major result of the interdisciplinary and cooperative work that Wright has fostered over his career. It has generated new understanding of global climate systems that can be applied to problems of global climate change. COHMAP is a classic example of the maxim that the past is the key to the present and the future, for it is data from the past that are needed to validate global models of climate. Our direct observations of climate span too little time and too narrow a range of factors for that purpose. Yet these models are our primary tool for projecting into the future the climatological effects of increasing greenhouse gases. Those effects, together with the social and political actions we humans take, will determine much of the character of the landscapes of the future.

These six chapters can present only a sampling of the areas of knowledge and expertise that contribute to reconstructing Quaternary landscapes and the processes that have changed them. Animals, though illustrated in our thematic landscape, are neglected, as are many geomorphic processes. There is bias toward high latitudes, too, for much remains to be learned about the history of tropical landscapes. Few of these areas have escaped Wright's attention, and he has furthered them as well through editing several books of contributions to Quaternary science. This slim volume can perhaps symbolize, if not introduce, those others.

The authors and editors hope that this book will encourage readers to fit together pieces of the landscape puzzle for themselves and to appreciate that the landscapes we live in, those environments that are both physical and cultural, have been determined by a rich and complex history. We hope our efforts will open some of the doors onto this multidisciplinary landscape to graduate and undergraduate students who are beginning to focus their attention on research problems. We hope it will help established researchers expand their horizons into areas close to their own fields. We feel that the chapters offer insight into the close relationship of many fields of scholarship, and that the book, like the man and institution it honors, provides a model to show the value of cooperative interdisciplinary studies.

Linda C. K. Shane and Edward J. Cushing
January 1991

Herbert E. Wright, Jr.

Herb Wright obtained his M.A. and Ph.D. degrees from Harvard University under Kirk Bryan. He has been a member of the faculty of the Department of Geology and Geophysics at the University of Minnesota since September 1947 and a Regents' Professor since 1974. He is an internationally known scholar involved with a wide range of ideas, people, and places. He has published over 150 articles, edited 14 books, directed 27 Ph.D. dissertations and 34 MA theses, and been involved with countless more. In North America he has conducted research across the United States and in Alaska, the Yukon, Labrador, and Mexico. Abroad he has worked in Lebanon, Iran, Iraq, Turkey, Greece, Norway, Sweden, Great Britain, Ireland, Peru, Bolivia, and Antarctica and made scholarly visits to the Soviet Union, China, Australia, and most European countries. He has received honorary degrees from Trinity College, Dublin, Ireland, and the University of Lund, Sweden, and numerous scientific awards, including membership in the National Academy of Science.

The focus of Wright's activity has been a desire (1) to reconstruct the Quaternary history of individual areas and ultimately of the world and (2) to use this reconstruction to increase our understanding of the present and the future. The word *history* here has a broad sense and includes geomorphological, climatic, biotic, and anthropological change. He has accomplished so much by carefully cultivating within the Limnological Research Center an atmosphere of cooperation, scholarly exchange, interdisciplinary research, and continuing international contacts. At this writing, the publication list for the LRC stands at 452 articles.

Wright has modeled much of his teaching and research style on that of his adviser Kirk Bryan. For his students and colleagues this has meant friendship combined with independence of thought and work, wide-ranging discussions, stimulating seminars in Herb and Rhea's home, and fantastic field trips with varied hardships. We have learned from him again and again that new technology such as isotopic dating techniques and computerized data manipulation provides only tools, not final answers; that scientific research is clearly and specifically founded in the most basic disciplines of orderly thought, constant questioning, and being sure of what one understands before moving toward speculation. In his leadership of the LRC, Herbert E. Wright, Jr., has shown us that careful research based on the desire for cooperation and discovery, and not tied to traditional scholarly boundaries, has both scientific and human value.

1
ORIGINS OF FOOD PRODUCTION IN WESTERN ASIA AND EASTERN NORTH AMERICA

A Consideration of Interdisciplinary Research in Anthropology and Archaeology

PATTY JO WATSON

Anthropology & Archaeology

When Robert J. Braidwood's Oriental Institute, University of Chicago, Iraq-Jarmo Project, left for its second field season in August 1954, I left with it. A very green, pre-M.A. graduate student straight from the Iowa cornfields, I was wildly excited at the prospect of nine months' archaeological research in an exotic part of the world known to me only from the *Arabian Nights* stories (expurgated version) and Biblical references to Palestine, Mesopotamia, the Assyrians, the Babylonians, the ancient Hebrews, and the Garden of Eden. I said I left with the expedition, but as a matter of fact, the other archaeological assistant—

Vivian Browman—and I missed the boat in New York City owing to a traffic jam in the Holland Tunnel. As we stood on the dock in a state of paralytic panic watching the *Exochorda* move slowly out into the harbor with our own baggage and most of the expedition gear on it (we were supposed to have been custodians of this shipment until disembarking in Beirut), an official ran up to us shouting "Andersen and Browman?!" "Yes!" we cried. He responded, "Follow me!" He ran us back across the dock, down some stairs, and onto a tugboat, which caught up with the *Exochorda,* aligned its deck to the big ship's open cargo door, and paused there while we were efficiently if unceremoniously tossed aboard by the willing hands of the tugboat crew. My career in interdisciplinary archaeology was launched.

The second, and as it turned out, the last season of the Iraq-Jarmo Project was the first full-scale, explicitly interdisciplinary archaeological team to be fielded in that part of the world since Pumpelly's Anau expedition of 1904. Braidwood's professional staff included a paleoethnobotanist (Hans Helbaek), a Pleistocene geomorphologist (Herbert Wright), a zoologist (Charles Reed), a ceramicist and radiocarbon specialist (Fred Matson), two archaeologists (Bruce Howe and Braidwood), and two graduate-student archaeological assistants (Vivian Browman and me). We were investigating the problem of the origins of food production in the Near East, the source of the economy that sustains Western civilization now as it sustained the first civilizations that arose in the Fertile Crescent of Mesopotamia and Egypt 5,000 years ago.

That memorable year of fieldwork significantly shaped my personal career, but it was not until much later that I realized all the implications of what I learned from Bob Braidwood, Hans Helbaek, Bruce Howe, Fred Matson, Charlie Reed, and Herb Wright. In particular, Herb's subtle, precept-and-example instruction was reinforced by several subsequent interdisciplinary fieldwork episodes with his crews and mine in both the Old World and the New.

Now, some 30 years after first meeting and working with him, I am delighted for the opportunity to dedicate to Herbert E. Wright, Jr., some thoughts on interdisciplinary research in general, and—in particular—interdisciplinary pursuit of knowledge about food-producing revolutions in the Near East and in eastern North America.

Development of the Interdisciplinary Approach in Anthropology and Archaeology

The development of interdisciplinary research in anthropology and anthropological archaeology has at least two major antecedents.

1. One is the 18th- and 19th-century European approach to the natural world, especially to the natural world in those exotic, faraway lands made known

to Europeans by their voyages of discovery during the 16th and 17th centuries and by their subsequent colonization efforts. The flora, fauna, and human inhabitants of these regions were described by European explorers and exploiters in a comprehensive manner familiar to us from natural histories of the 18th, 19th, and early 20th centuries. Physical appearance, dress, manners, and customs of the human natives were detailed together with the geography, geology and plant and nonhuman animal life. Uncivilized (i.e., non-European and non-Christian) human groups were viewed as part of the natural world. A souvenir of this perspective is clearly evident in many of the natural-history museums in the contemporary United States. The Smithsonian National Museum of Natural History in Washington, D.C., the Field Museum of Natural History in Chicago, and the American Museum of Natural History in New York, for example, all divide natural history into anthropological, botanical, geological, and zoological quadrants.

2. Similarly, the growth of geology and paleontology in the 17th, 18th, and early 19th centuries in Europe had a direct effect on the early development of prehistoric archaeology there and in North America. Battles about the antiquity of the earth were paralleled by those about the antiquity of humankind on earth (Daniel, 1975; Grayson, 1983; Heizer, 1962; Lyell, 1830–1833). By the later 19th century, in the wake of Lyell and Darwin, European-based anthropology grew in part as the study of lifeways illustrative of early stages in human cultural and social evolution. Thus, 19th-century biology as well as geology and paleontology heavily influenced early anthropology and prehistoric archaeology—paleolithic archaeology being especially closely tied to the earth sciences.

In North America, archaeology was a branch of anthropology from the beginning of its development; studies of ancient and living Indians were not separated (Taylor, 1948: 21). For example, the first formal ethnography in the United States was produced by Frank Hamilton Cushing, who was sent to Zuni Pueblo in New Mexico by the brand-new Bureau of Ethnology in 1879. Cushing learned the language, studied the Zuni for four years, and published a number of ethnographic accounts, but he also made archaeological observations and carried out excavations in the Zuni area as a natural extension of his research on the contemporary pueblo people (Green, 1979).

In the early to mid-20th century, several Americanist archaeologists advocated a holistic approach to obtaining anthropological knowledge from the archaeological record, which they viewed as a corpus of paleoethnographic data. The most prominent and influential include Alfred Kidder (1924, 1937), Walter W. Taylor (1948), and Gordon Willey and Philip Phillips (1958). For several decades, Willey and Phillips's remark (1958: 2), "Archaeology is anthropology or it is nothing," accurately expressed the professional attitude of archaeologists trained in prehistoric archaeology in the United States.

In the Old World, interdisciplinary research in archaeology was pioneered by

Raphael Pumpelly at Anau in central Asia (Pumpelly, 1908). Pumpelly was a geologist, having spent "three-score years and ten," as he says, practicing that profession in China and Mongolia before turning to archaeology. Hence, it is not surprising that his field staff for the 1904 season at Anau included a geomorphologist (his son, R. Welles Pumpelly) and a cultural geographer (Ellsworth Huntington) as well as four archaeologists (Hubert Schmidt, Hildegard Brooks, Homer Kidder, and Langdon Warner). The field staff put in long hours. Pumpelly notes that work began at 4 a.m., stopped for one hour at breakfast time and two hours at lunch time, but did not cease for the day until 6 p.m. When a plague of grasshoppers stopped the work at Anau, they continued at Merv and Samarkand. The materials recovered from those places were studied and published by a zoologist (J. Ulrich Duerst), a chemist (F. A. Gooch, who analyzed the metal artifacts), two human paleontologists or physical anthropologists (G. Servi and Th. Mollison, who described human remains from a burial mound), a botanist (H. C. Schellenberg, who identified wheat and barley from the burial mound), and three of the archaeologists (Homer Kidder, Hubert Schmidt, and Langdon Warner).

Pumpelly's example of successful interdisciplinary archaeological research was not followed, however, until the middle of this century in the years after World War II. During the 1950s, interdisciplinary, ecologically oriented archaeology with paleoenvironmental reconstruction as a primary goal took on its modern form. Two names come immediately to mind when one thinks about the early development of this approach in the early postwar period: Grahame Clark of Cambridge University for his 1952 book, *Prehistoric Europe: The Economic Basis,* and his interdisciplinary research at the Mesolithic site of Star Carr; and Robert J. Braidwood, Oriental Institute and University of Chicago, who fielded the first interdisciplinary teams to search for agricultural origins in western Asia (Braidwood, 1960; Braidwood and Howe, eds., 1960).

Braidwood's Iraq-Jarmo project was funded by the National Science Foundation and was enormously influential in both Old and New World prehistoric archaeology. In Mesoamerica, MacNeish's Tehuacan Project (Byers, ed., 1967; MacNeish, 1967) was the first of many explicitly interdisciplinary archaeological investigations following the Braidwood model. This paleoenvironmental, paleoecological, interdisciplinary focus for archaeology blossomed during the 1960s and 1970s as a central component of Americanist New Archaeology (Watson *et al.*, 1971, 1984). Sometimes called "processual archaeology," its proponents were also strongly committed to the view that archaeology is best practiced as a social science (Binford, 1962). In England, a separate but similar formulation was eloquently propounded by David Clarke (1968), who stressed (rather than a social science approach per se) the great potential of systems theory and spatial analysis for archaeologists.

At the present time, Americanist archaeology continues to be dominated by the interdisciplinary, paleoecological approach, which has also found some acceptance within the very different disciplines of biblical and classical archaeology (Dever, 1981; Dyson, 1981; McDonald and Rapp, eds., 1972). But a number of new trends and old problems are increasingly prominent; these require the serious attention of ecologically oriented, archaeological interdisciplinarians.

Problems and Challenges

Fieldwork Problems

By definition, successful interdisciplinary research requires an inordinate amount of careful coordination and detailed intercommunication. These matters are even more critical in a discipline like archaeology where a large component of the research is fieldwork, traditionally in physical, social, and political circumstances that are usually strenuous and sometimes dangerous. These challenges to interdisciplinary archaeology have not always been met successfully (Butzer, 1975, 1978, 1980; Schoenwetter, 1981). Even where the fieldwork is well managed, there are likely to be delays and serious disagreements among the personnel about completing analyses and publishing results (Watson et al., 1984: 232–42). Until very recently, few archaeologists were trained as professional administrators and tension-managers, yet such skills are essential to the design, implementation, and publication of interdisciplinary research.

Contract/CRM Archaeology

A new development in the United States has had significant impact on these older problems inherent to interdisciplinary archaeology, and that is the creation in 1974 of contract or cultural resource management (CRM) archaeology. Federal legislation requiring assessment of historic and prehistoric materials to be affected by federally funded construction of any sort marked the beginning of a new era for archaeologists working in the United States (Dunnell, 1979, 1980, 1981, 1982, 1983, 1984, 1985; King *et al.*, 1977; McGimsey, 1972; Schiffer and Gumerman, eds., 1977). Discussion of the early history and current status of CRM is far beyond the scope of this chapter, but one point is important here. The availability of relatively generous budgets meant that interdisciplinary archaeology—always very expensive and subject to the vagaries of bootlegged time and ad hoc arrangements with collaborators—could be straightforwardly funded in a way seldom possible previously, even with a large NSF grant. This fact has been used to very good advantage by some archaeologists (e.g., Bense, ed., 1983; Bareis and Porter, eds., 1984; Chapman, 1985; Faulkner and McCollough, eds., 1973, 1982).

POST-PROCESSUAL ARCHAEOLOGY

The other recent development with important implications for interdisciplinary, anthropological archaeology is very different from the birth and growth of CRM in the United States. It is nothing less than a global intellectual challenge to contemporary archaeological theory and practice, referred to by one of its chief architects, Ian Hodder, as "post-processual archaeology." By this phrase, Hodder means to imply and to advocate a broadening of archaeological endeavor so that it includes "Marxism, structuralism, idealism, feminist critiques and public archaeology. At the same time, the aim is to establish archaeology as a discipline able to contribute an independent voice to both intellectual and public debates" (Hodder, 1986: 171). Hodder and his supporters have published their views extensively. It is sufficient for present purposes to state that Hodder is primarily concerned with the *meanings* that archaeological materials had for their creators and users, and with the more general issue of meaning in the production and functioning of material culture, past or present. He thinks the goal of archaeology should be to attain a comprehensive understanding of the cognitive systems of past societies. These systems formed the contexts in which the materials we perceive as the archaeological record were purposefully created, manipulated, lost, and worn out or destroyed. Thus, the ultimate explanation of archaeological remains is the meanings they embody. If analyzed appropriately, they can still convey information to us about those original social and cultural contexts. Although Hodder does not deny the necessity of economic and materialist analyses, he believes them to be restrictive and not very interesting because he sees no necessary relationship between social and material organization of resources on the one hand, and cultural ideas and values on the other. Because his focus is exclusively on the latter, his formulations give short shrift to the ecologically oriented, adaptive concerns of most recent archaeological research.

Hodder's views have been widely publicized and have been sympathetically received in some places—especially England and parts of the European continent—but have been rejected in others. Debate and discussion about Hodder's program continues among Americanist archaeologists, though the great majority go on pursuing paleoecological/paleoenvironmental questions about prehistory (Watson, 1986, 1990). It seems unlikely that post-processualism will rise to a position of dominance in North American archaeology in the near future, but the narrow focus on materialist, economic, and paleoenvironmental issues will be widened sufficiently to allow analyses and concerns representing some of the perspectives advocated by Hodder and others. Indeed, such a trend has been under way for some time (e.g., Bender, 1985a, 1985b; Braun, 1977, 1988; Brown, 1985; Conkey and Spector, 1984; Gero, 1985; Gero and Conkey, Eds., 1991; Gero *et al.*, eds., 1983; Leone, 1978, 1982, 1986; Marquardt, 1985; Prentice, 1986).

The remainder of this chapter illustrates some of the themes just noted while discussing the current status and results of the interdisciplinary approach as applied to one major problem in two areas of the world where Herb Wright's paleoenvironmental research has been critical and foundational: the origins of food production in western Asia and in the Eastern Woodlands of North America.

The Origins of Food Production in Western Asia and Eastern North America

In considering this perennially fascinating topic, I refer first to the empirical data or evidential base for each region—that is, what is known—then to the question of interpretation of those data, how they are understood or explained. Although such a separation can be made analytically, or for purposes of organizing the discussion, data and interpretation are, of course, inextricably entwined in archaeology, as they are in any other empirical discipline.

Western Asia: The Data

Several recent summaries of the relevant materials from western Asia are available to supplement and update older syntheses (Braidwood and Braidwood, 1986; Henry, 1983, 1985, 1989; Mellaart, 1975; Miller, in press; Moore, 1985; Redman, 1978), yet the current pace of work in the Levant and in several dam-salvage regions in southern Turkey, northern Syria, and northern Iraq means that anything in print is likely to be somewhat out of date. Even so, a number of important generalizations can be made.

First, largely as a result of research initiated by Herbert E. Wright and Willem van Zeist in Iran in the 1960s (van Zeist and Wright, 1963; van Zeist, 1967), the late Pleistocene and early to middle Holocene vegetation for large portions of the Near East can be described in increasing detail (Bintliff and van Zeist, eds., 1982; van Zeist and Bottema, 1982, unpublished data; Wright, 1976b, 1977, 1983). Climatic history inferred from pollen sequences (in some places with support from faunal and geological studies; see Henry [1989] for a useful summary) is clear in outline, though minor fluctuations and detailed regional variations are still being investigated. Except for the southern Levant, the Near East was cold and dry during the final portions of the last glaciation. Cold, dry steppe conditions prevailed in the interior (with alpine conditions in the high mountains) until ca. 11,000 yr B.P. *Artemisia* (wormwood) and *Chenopodium* (lamb's-quarters and goosefoot) pollen dominate the sequence, and arboreal pollen is rare to absent. After 11,000 yr B.P., warmer and moister conditions developed, allowing Mediterranean oak-savanna vegetation to expand from various refuges (primarily in the central and southern Levant) and sheltered areas until by approximately 5,000–4,000 yr B.P. the mountains and foothills were forested, and the lowlands were warm-steppic in character.

In the southern Levant, trends and fluctuations were more complex (van Zeist and Bottema, 1982: 283–84; Henry, 1983: 106–14, 1989: 64–78). The pollen records indicate dry, cold steppe at about 24,000–14,000 yr B.P.; oak woodland expanded between 14,000 and 10,000 yr B.P. but decreased between 10,000 and 7,400 yr B.P. After 7,400 yr B.P. forest expanded once more and the present vegetation regime was established. Some pollen information for (presumably localized) conditions in northern Syria suggests that the climate at approximately 10,000 yr B.P. was moister than it is now (see the Abu Hureyra-Mureybat discussion below).

Thus, evidence indicates marked regional variation in climatic regimes during the late Pleistocene and early to middle Holocene in the Near East. Before considering the implications of the paleoenvironmental picture, however, we must take some account of the archaeological information for the period before and during the first appearance of domesticated plants and animals.

The horizons in question are variously called the Mesolithic, the Epipaleolithic, or the eras of terminal food collecting and of incipient food production; and the Archaic Neolithic, the Aceramic or Prepottery Neolithic (PPN), or the era of primary, village farming communities. A simpler terminology is to refer to the following kinds of economies or subsistence systems: *foraging* (unspecified gathering and hunting of a variety of wild flora and fauna); *foraging and harvesting* (the same as foraging, but with special attention to a few particularly abundant or culturally favored plant or animal resources); *early horticulture* (the cultivation of at least a few plant species to the extent that they are eventually, morphologically or otherwise, recognizable as domestic); *early herding* (tending one or more animal species and monitoring and manipulating their life cycles to the extent that they are eventually recognizable morphologically or by age distributions and other statistical means as domestic); *agriculture* (heavy reliance on domestic plants to the virtual exclusion of wild species, at least insofar as staple foods are concerned); and *pastoralism* (heavy reliance on domestic animals to the virtual exclusion of wild species, except for special purposes such as ritual activity, furs, or sport).

Archaeological evidence is slim for the middle Pleistocene in the Near East, and vanishingly rare for the early Pleistocene, but it is most likely that all the human groups scattered over western Asia during those periods were foragers. Near the end of the late Pleistocene, however, and in the beginning of the early Holocene, some groups in the Zagros and in the Levant had apparently developed harvesting economies with varying emphases on wild cereals and legumes and perhaps on gazelle, deer, goat, or sheep populations.

Archaeological manifestations of these harvesting and foraging societies are known as Kebaran in the central and northern Levant, Hamran in southern Jordan, both succeeded by the Natufian before 12,000 yr B.P. (Bar-Yosef, 1980; Henry, 1989: 119; Moore, 1985). In the Zagros they are known as the Zarzian, Zawi Chemian, and Karim Shahirian (Braidwood and Howe, eds., 1960; Braidwood *et al.*, eds., 1983; Solecki, 1981; Turnbull and Reed, 1974). The sites are rock shelters and small open-air settlements of round, semisubterranean houses and storage structures with abundant and varied stone-tool industries including grinding-stone complexes (mortars and pestles, querns and handstones) as well as celts and chisels. Bone tools are also common. At several sites human burials have been found. Excavators and interpreters of some of the open sites regard them as representing semisedentary, or "relatively permanent" (Henry, 1983: 138, 1989: ch. 7) hamlets and base camps whose inhabitants subsisted on diverse wild plant and animal resources, but emphasized cereals, nuts, gazelle, roe and fallow deer, goat, and sheep. At one of the Zagros rock shelters, for example (the Zarzian site of Palegawra in northern Iraq), nearly 775 of the faunal remains were from only two genera (*Equus* and *Cervus,* onager and red deer), whereas at two other Zagros sites, Shanidar and Zawi Chemi Shanidar, the faunal sequence has been interpreted as indicating the earliest evidence (just after 11,000 yr B.P.) of at least incipiently domestic sheep (Perkins, 1964). This specific interpretation has been resisted by some (Ducos and Helmer, 1981; Wright and Miller, 1976), but it is theoretically plausible (see also Evins, 1983, and Hole, 1984).

Early horticultural societies raising wheat, barley, and legumes appear in the archaeological record ca. 10,000–9,500 yr B.P. (8,000–7,500 b.c.e.) at Jericho, Nahal Oren, and Tell Aswad, all in the Levant; Çayönü in southeastern Turkey; and at Ganj Dareh and Ali Kosh in the Zagros (Braidwood and Braidwood, 1982; Braidwood *et al.*, 1981; Clutton-Brock, 1979; Helbaek, 1969; Hopf, 1969; Miller, in press; Moore, 1975, 1985; Noy *et al.*, 1973; Smith, 1978; van Zeist, 1972; van Zeist and Bakker-Heeres, 1979, 1982 [1985], 1984 [1986], 1985 [1988]; van Zeist *et al.*, 1986). Some of these sites also provide evidence for domestic sheep and goats at the same time as the early plant domesticates (Ali Kosh, for example), or in later levels than the domesticated plants (e.g., Çayönü).

Two sites in Syria merit special attention, Abu Hureyra and Mureybat. These mounds are on the Euphrates in northern Syria and are less than 40 km apart; Mureybat is on the east bank and north of Abu Hureyra, which is on the west bank. The basal deposits at Abu Hureyra (referred to originally as Mesolithic [Moore, 1975, 1985], and now as Epipaleolithic [Hillman, 1988; Hillman *et al.*, 1986; Legge, 1988; Molleson, 1988; Moore, 1988; Olsen, 1988; Olszewski, 1988; Rowley-Conwy, 1988]), date to 11,000–10,100 yr B.P. (ca. 9,000–8,100 b.c.e.) and are thus older than the rest of the sequence, which runs from about 9,600 to 7,500 yr B.P. (7,600 to 5,500 b.c.e.), partially overlapping the Mureybat deposits, which begin at about 10,500 yr B.P. (8,500 b.c.e.) and end just after 9,000 yr B.P. (7,000 b.c.e.). There are remains of einkorn wheat, rye, lentil, vetch (probably all wild), and a diverse array of weedy species in the earliest Abu Hureyra levels, and of domestic wheat (both emmer and einkorn), domestic 2-row and 6-row barley, and domestic rye in the later (10th millennium B.P./8th millennium b.c.e.) levels (Hillman, 1975, 1988; Hillman *et al.*, 1986). Wild-type einkorn wheat, barley, and lentils are also present at the 11th millennium B.P. (9th millennium b.c.e.) at Mureybat (Cauvin, 1977; van Zeist and Bakker-Heeres, 1984 [1986]; van Zeist and Casparie, 1968), but their status—wild, or morphologically wild but cultivated—is unclear. Although Leroi-Gourhan, (who analyzed pollen from Mureybat) suggests that the quantities and size of cereal pollen grains in the 9,700 yr B.P. (7,700 b.c.e.) and later levels may indicate that "proto-agriculture" was under way there at that time (Leroi-Gourhan, 1974), van Zeist and Bakker-Heeres (1984 [1986]: 194–99) think this unlikely.

Thus, both sites contain deposits of the crucial time period (11th millennium B.P./9th millennium b.c.e.), and both have yielded considerable archaeobotanical material, but the data are ambiguous for two reasons.

1. There is a gap in the stratigraphy at Abu Hureyra from ca. 10,100 to 9,600 yr B.P. (8,100 to 7,600 b.c.e.). The deposits earlier than the gap contain wild plant species only (although einkorn wheat and rye are among them), and the deposits that follow the gap contain a complete set of domestic cereals: emmer and einkorn, 2- and 6-row barley, and domestic rye.

2. The deposits at Mureybat for the critical period of the Abu Hureyra gap (10,100 to 9,600 yr B.P. or 8,100 to 7,600 b.c.e.) contain abundant plant remains including small quantities of einkorn, barley, and lentils, but these are morphologically wild, as are the remains of these same species from later levels at the site.

The simplest explanation of these data is that the 11th millennium B.P. (9th millennium b.c.e.) inhabitants of Abu Hureyra and Mureybat were foragers and harvesters living a sedentary or semisedentary existence on the basis of a number of wild plants and animal species, including wild einkorn. As noted above, there is some pollen and macrobotanical evidence to the effect that the local climate was moister ca. 10,000 yr B.P. than now (see Hillman *et al.*, 1986; Niklewski and van Zeist, 1970; van Zeist and Bakker-Heeres, 1984 [1986]: 194–99; van Zeist and Woldring, 1980). Therefore, wild einkorn probably grew closer to these sites at that time than it now does, when it is found no nearer than 100–150 km north.

As the local environment became more arid in the early to mid-10th millennium B.P. (8th millennium b.c.e.), a complex of domesticated plants (those listed in point 1 above) and animals (probably sheep and goat, but the evidence is not entirely straightforward—Legge, 1988; Rowley-Conwy, 1988) was introduced from the west, where domestic grains (emmer, einkorn, perhaps *Triticum durum*, 2-row barley) and legumes are known from early 10th millennium

B.P. contexts at Tell Aswad near Damascus (van Zeist and Bakker-Heeres, 1985 [1988]) and from Prepottery Neolithic A Jericho and Nahal Oren (Hopf, 1969; Noy *et al.*, 1973). This agricultural and pastoral complex is visible archaeologically at Abu Hureyra in the mid- to late 10th millennium B.P. levels, but is either absent from Mureybat or has not yet been detected there. If absent (perhaps more likely as the site has been excavated on two different occasions by two different archaeologists), then Mureybat is—as is basal Abu Hureyra—a settlement of sedentary foragers and harvesters, as van Loon originally described it (1968).

A harvesting economy may have been viable at the more northerly site of Mureybat when it was no longer adequate at Abu Hureyra. Mureybat is not only a little farther north than (hence upstream of) Abu Hureyra, but also it is on the east side of the Euphrates, and thus closer to the low hills that rise above it between the Balikh River on the east and the Euphrates to the west before the Euphrates makes a big easterly bend in its descent toward north Iraq. Perhaps this local situation at Mureybat was moister than that at Abu Hureyra, which is on the south side of the Euphrates just below the big bend. The preliminary discussion by van Zeist and Bakker-Heeres (1984 [1986]: 197) of two differences in the plant remains at the two sites might be commensurate with a suggestion of drier conditions at Abu Hureyra than at Mureybat: absence at Mureybat but abundance at Abu Hureyra of two plants characteristic of dry steppe conditions—*Onobrychis* and *Stipa*. There are many other steppic plants at Mureybat, however, and van Zeist and Bakker-Heeres (1984 [1986]) state that most of the plants known from Mureybat could have formed part of a steppe community, which they suggest was present on the uplands above the floodplain where Mureybat is located. So the evidence is far from conclusive.

As Leroi-Gourhan (1974) suggests, some form of plant cultivation may have been adopted in the last stages of the occupational sequence at Mureybat, when domestic plants and animals were also introduced into Abu Hureyra (perhaps the Abu Hureyra rye was a local domesticate, however). There are of course other possible, nonclimatological explanations for the absence of *Onobrychis* and *Stipa* at Mureybat (accidents of preservation and sampling, differential activities by the prehistoric occupants of the two sites, and so on).

Other than Zawi Chemi Shanidar (see above), the only eastern site with relevant biological remains that may be as old as the 11th millennium B.P. (9th millennium b.c.e.) is Ganj Dareh in the Iranian Zagros (Smith, 1978; van Zeist *et al.*, 1986). Here, as at Abu Hureyra, the basal occupation of stratigraphic level E (a seasonal camp?) is thought to be pre-10,000 yr B.P. (8,000 b.c.e.). In E as well as in the later portion of the sequence (horizons D to A, mid-10th to mid-9th millennium B.P./mid-8th to mid-7th millennium b.c.e.), wild and domestic 2-row barley are present (but no wheat), and there were probably domestic goats (Hesse, 1982). By this time the settlement was apparently occupied all year round (Hesse, 1979).

Also in the Iranian Zagros Mountains in the site of Ali Kosh, domestic einkorn, emmer, and barley are reported for the basal occupation (Bus Mordeh phase, 10th millennium B.P./8th millennium b.c.e.), and sheep and goats were also herded (Helbaek, 1969; Hole *et al.*, 1969). At Cayönü in southeastern Turkey, the earliest levels (10th millennium B.P.) contain domestic einkorn, emmer, and (probably domestic) lentils and peas (but no domestic sheep or goats until the later, 9th millennium B.P./7th millennium b.c.e. deposits).

The Reasons Why

As the above précis should make clear, data bearing directly on the development of horticultural, herding, agricultural, and pastoral economies from earlier foraging and harvesting subsistence systems are sparse and scattered for the Near East. So it is not surprising that interpretations and explanations of these data are not entirely congruent. Although a comparative summary of recent opinions is provided here, it is unlikely that a detailed consensus will emerge until much more is known about the specific requirements and behavior of the wild faunal and floral populations in question (especially the plants) and until there is a much broader and stronger body of archaeological and ecofactual evidence for the 11th and 12th millennia B.P. (9th and 10th millennia b.c.e.).

The major factors for all discussants remain the same, however: the nature and effect (1) of environmental change, (2) of population increase, and (3) of sedentism. Many writers on the subject of agricultural and pastoral origins in the Near East project a monolithic perspective in discussing agriculture (or even proto- or incipient agriculture) and pastoralism as if these were discrete, clear-cut subsistence pursuits, readily identifiable and readily transferable from one place to another, more or less en bloc. As more detailed information about the harvesting, early horticultural, and early herding economies of the Near East is gained, this en bloc approach will disappear because an intricate mosaic of subsistence patterns will surely be revealed.

A comparison of Henry's (1983, 1985, 1989) and Moore's (1985) recent interpretations affords a useful illustration of the variation in contemporary views (see Moore, 1985: 43–49, for an account of older formulations from the 1930s to the late 1970s; also see parenthetical comment for Moore [1988] below).

Henry's interpretation (Henry, 1983: 150–51, 1985, and 1989) starts with the well-documented late Pleistocene–early Holocene climatic change and then takes up the postulated interplay between certain natural resources (primarily cereal grains), human subsistence behavior, and human population growth. The sequence he proposes is as follows. At approximately 12,500 yr B.P., the Mediterranean woodland (including the cereal species that are part of this ecological community) expanded from its late Pleistocene refuges, resulting in a new subsistence focus (on barley and emmer; acorns, almonds, and pistachios) for the inhabitants of the Levant (the Natufians) and probably also for the human populations living in the Zagros at places like Zawi Chemi Shanidar. Henry believes that stands of wild emmer and barley, once they had colonized the Mediterranean hill zone, were much denser, more vigorous, and more predictable in their distributions year after year than they would have been in the pre-Holocene refuge areas, and they would therefore have emerged as a very appealing resource for human foragers. Intensive exploitation of them, however, would have required year-round attention to their growth, and to the processing and storage of the harvest, so that fixed settlements (sedentism) would have become advantageous. Population increase, then, follows sedentism, and significant innovations in social organization (from simple to complex hunting-gathering societies in his terminology; Henry, 1985) follow population increase, up to and including the appearance of social ranking and chiefdomlike societies (Henry, 1983: 142–45). Henry (1983) does not discuss the role of animal resources specifically, nor does he continue his formulation beyond this point, but leaves prehistoric Levantine society at 10,000 yr B.P. intensively exploiting wild cereals in a state of preagricultural sedentism. In later publications (Henry, 1985, 1989), he does discuss the probable consequences for these dense and widely distributed, complex foraging and harvesting Natufian groups when they were confronted by a reversal (ca. 11,000 B.P.) of the moist climatic conditions that originally encouraged expansion of Mediterranean woodlands throughout the Levant between 13,000 and 11,000 yr B.P. Reduction of the Mediterranean flora (including the cereal habitats, presumably) after 11,000 yr B.P. must have resulted in an imbalance between human populations and the vital plant resources they had been accustomed to rely on. Regionally, the Mediterranean vegetation retreated north and west as precipitation declined; locally, biotic zones moved upslope (or disappeared altogether). As a result, only those human communities with access to permanent water sources—for example, Nahal Oren and Jericho—could adjust by experimenting with plant cultivation to reproduce earlier conditions when stands of wild grain were available nearby. Thus, Henry (1985, 1989) views the development of domestic grains as an attempt by some complex foraging and harvesting groups to perpetuate that lifeway against increasing environmental odds. Other complex groups, however, seem to have turned to the more dispersed and mobile lives of simple foragers.

Andrew Moore (1985) takes up the story at this stage and carries it to the appearance of well-established agriculture and pastoralism. (As this paper was going to press, I learned [Moore, 1988] that new ^{14}C dates by accelerator mass spectrometer and new analyses have resulted in modification of Moore's views as provided in his 1985 synthesis. The new interpretations will be available soon when the eagerly awaited Abu Hureyra report is published.) A point of agreement between the two authors is the issue of population increase. Moore and Henry both conclude that human densities increased considerably in the 12,000 to 10,500 yr B.P. time period over previous levels. Moore notes the appearance of sedentism at this same time, but seems to attribute it primarily to reduction in size of hunting territories as the terrain became more crowded with human settlements (Moore, 1985). He suggests that the first deliberate planting and harvesting of cereals in the Levant was done by people living in what were at that time relatively well-watered, steppe-edge situations like that of Abu Hureyra, whereas the contemporary Mediterranean forest dwellers—the Natufians, for example—were supported more by nuts than by cereals. (He clearly disagrees with Henry here.)

Moore's 1985 construction begins, like Henry's, with the environmental change that commenced about 13,000 yr B.P. when temperature and humidity increased, enabling expansion of the forest zone. Growth of human populations followed this improvement in the physical environment, resulting in reduced hunting territories and greater concentration on newly abundant plant and animal resources. Sedentism ensued, and also coalescence or agglomeration into villages, some of which were occupied permanently. A higher rate of population growth resulted from prolonged sedentism, increasing pressure on local wild plant and animal foods to the point where they had to be systematically manipulated and exploited. This set of processes culminated in the domestication of both plants and animals and the emergence of horticultural and herding societies during the 12th millennium B.P. (10th millennium b.c.e.) in at least a few places in the Levant. Full-scale agricultural and pastoral economies developed during the 11th and 10th millennia B.P. (9th and 8th millennia b.c.e.), and soon after 8,000 yr B.P. (6,000 b.c.e.), subsistence systems in the Mediterranean woodland zone were focused on domestic plants and animals to the virtual exclusion of wild species. Sedentary village life and the beginning of agriculture were broadly synchronous, and there was no significant social differentiation until after 6,000 yr B.P. (4,000 b.c.e.), that is, the earlier (Epipaleolithic in his terminology) foraging-harvesting and later (Neolithic) horticultural and herding societies were egalitarian. On both these points, Moore (1985) differs significantly from Henry, who believes stable sedentism long precedes food production, and that ranked society was present in the Natufian.

In a recent discussion of early agriculture and pastoralism in the Near East, Hole (1984) stresses the differential development of plant and animal domestication. He notes evidence for an early center (late 11th millennium B.P./9th millennium b.c.e.) of goat and sheep domestication in the Zagros (e.g., at Zawi Chemi Shanidar, Ganj Dareh, and Ali Kosh), though the ultimate origin point for domestic sheep (and perhaps goats, too) could be central Asia. Cattle were probably domesticated first in Anatolia (by the mid- to late 9th millennium B.P./7th millennium b.c.e.); domestic pigs—also present by the mid- to late

9th millennium B.P.—are much more localized in importance than are the other three species and are thought to be poorly adapted to fully nomadic or to transhumant pastoralism. Domestic food animals were introduced into the Levant in the 9th millennium B.P. as indicated at several PPN B sites, such as Jericho, Abu Gosh, and Beisamoun. Hole contrasts this pattern of early animal domestication with the information about plant use, which indicates significant dependence on cereals in arid regions beginning in the 12th millennium B.P. (10th millennium b.c.e.), and clear evidence from the Euphrates to the Mediterranean for domesticated species by the 10th millennium B.P. (8th millennium b.c.e.) in the absence of any domesticated ungulates.

Hole suggests a two-stage sequence for the food-producing revolution with the differential domestication trajectories for plants and animals, as just noted, comprising stage one; stage two comprises the convergence of plant and animal domesticates (at, e.g., the Levantine PPN B sites, Çayönü, Ganj Dareh D, and basal Ali Kosh). Hole also notes that social organizational and ideological factors must have been crucial in the stage one processes that led to the lifeways significantly dependent on domestic species. He suggests that sedentism, population increase, and other fruits of the Neolithic Revolution resulted from the "multiplier effect" of linking economies based on domesticated plants with those featuring domestic food animals.

In concluding this overview of the origins of food production in the Near East, one should take note of two other less-detailed but intriguing and contrasting interpretations, both somewhat more post-processual in nature: those of Bender and Cauvin.

Bender (1978) believes that the adoption of an agricultural economy happens because of changing social relations, that is, social relations are paramount and they, rather than environmental change or technological innovations, function as the independent variable. She notes further that "the enquiry into agricultural origins is not, therefore, about intensification per se, not about increased productivity, but about increased production and about why increased demands are made on the economy" (1978: 206). Bender finds considerable fault with recent attempts to explain agricultural origins by evoking demographic pressure as a prime mover (she is referring primarily to Binford [1968] and Cohen [1977]). She points out, among other things, that sedentism is not necessarily an easy option socially because mobile hunting-gathering (foraging) groups have few means for controlling intragroup conflict other than fissioning. There are also significant labor costs involved in building and maintaining permanent dwellings and storage facilities, and serious health hazards in sedentary life that are unknown or very rare for mobile societies. Finally, she says that demographic pressure is not a matter of absolute population density in relation to natural-resource potential; rather it is culturally defined and is noticed only when the social status quo is clearly threatened.

In her alternative formulation Bender stresses what has been demonstrated ethnographically about the importance of reciprocity and alliances among the subgroups (bands, lineages, and residential units) of hunting-gathering societies. She then shows how individuals cooperating and competing within systems of reciprocal trade, marriage alliance, and exchange can cause, or at least encourage, surplus production (intensified harvesting) of wild resources, especially foodstuffs; sedentism; and eventually the intensification of subsistence to the point of food production based on domestic species.

Cauvin (1977, 1978) takes a somewhat similar tack to that of Bender in discussing the archaeological remains at Tell Mureybat. He suggests that the prehistoric Mureybat population took up agriculture in the context of tensions resulting from population growth, and that this economic change was deliberately made to resolve or ameliorate social conflict. Like Bender, he stresses the importance of individual choices and actions, and thus diverges sharply from the ecological, paleoenvironmentally structured interpretations of Henry (1983) and Moore (1985), which are cast far above the level of individual human beings and their behavior (in another paper, Henry [1985] does include discussion of individual decision making).

Both these categories of explanation—ecological/paleoenvironmental and social relational (quasi- or semi- post-processual)—are also present in the literature about the development of horticulture in the Eastern Woodlands of North America, as indicated below. First, however, some attention must be given to the pertinent portion of the archaeological record there.

EASTERN NORTH AMERICA: THE DATA

The issue of horticultural origins in the Eastern Woodlands of North America is an extremely lively one at the present time. In the decade and a half since flotation techniques became standard (Hastorf and Popper, eds., 1989; Struever, 1968; Wagner, 1982, 1989; Watson, 1976), the database has expanded rapidly and continues to do so. Fortunately, there are current summaries of the available information on plant use and cultigen development (Asch and Asch, 1985; Heiser, 1985; King, 1985; Smith, 1987a, 1989, in press; Watson, 1985, 1988, 1989; Yarnell, 1983, 1986) that enable me to be quite succinct in outlining the evidence to date. Before taking up that topic, however, it is necessary to turn to the issue of paleoenvironment.

The time period in question is the middle Holocene and initial late Holocene, rather than the late Pleistocene and early Holocene as in western Asia. Once again many of the crucial data on paleoclimate and vegetation derive from the research of H. E. Wright, his students, and his collaborators (e.g. Delcourt and Delcourt, 1979, 1981, 1983; Jacobson *et al.*, 1987; King, 1981; Ruddiman and Wright, eds., 1987; Watts, 1980, 1983; Webb *et al.*, 1983; Whitehead,

1973; Wright, 1976a, 1981; Wright, ed., 1983). As in the Near East, the most dramatic environmental aspect of the postglacial period is the appearance and expansion of those arboreal species absent or distributionally much reduced during the full glacial regime. Because there was so much variation latitudinally in the eastern United States, the Delcourts (1983) have divided this area into three zones from north to south: north of 43°N latitude (43°N is approximately the Canadian border), 43° to 34°N latitude (approximately from the Canadian border to the latitude of Charlotte, North Carolina), and 33° to 29°30′N (approximately from Columbia, South Carolina, to St. Augustine, Florida). The archaeological materials pertaining to the beginnings of plant cultivation derive from the middle and the most southerly of these zones. Vegetation of the Gulf coastal plain in the southernmost zone, being under the influence of tropical, maritime conditions, has been fairly stable over the past 14,000 years, featuring oak, hickory, and southern pine as dominants in the uplands above the coast. Elsewhere in the southern zone, mixed hardwood species grew along the larger rivers, forming a more mesic woodland (oak, hickory, tulip poplar, beech, maple, walnut, elm, basswood) that expanded north after the glacial retreat, ultimately reaching the Appalachian Mountains.

The Floridian peninsula was characterized by nonarboreal vegetation and active dune fields during the full-glacial time, replaced by oak savanna and ultimately by southern pine as the climate ameliorated. The pine forest also expanded along the coastal plain to the north.

At about 10,000 yr B.P. prairie, oak savanna, and oak-hickory woodland comprised the vegetation of the middle zone from west to east. A deciduous forest dominated by oak-hickory (with beech and some hemlock) was present in the southern portion of the zone; pine was present to the north; elm and ash (plus some ironwood) were present to the west, extending into the eastern edge of the present Great Plains (Jacobson *et al.*, 1987). During the mid-Holocene (approximately 8,000 to 5,000 yr B.P.), warm, dry westerlies caused a decrease in summer rain, resulting in eastward expansion of oak savanna and prairie communities (see King, 1981, for a detailed summary of vegetation history in a portion of the Midwest with special reference to the Prairie Peninsula that covered large parts of Iowa, Missouri, and Illinois beginning ca. 8,000 yr B.P.). This warm, dry episode is known as the Hypsithermal interval. By ca. 5,000 yr B.P., the vegetation of the Midwest, the Midsouth, and the Southeast had assumed its modern, postglacial character.

The oldest archaeological remains in the Eastern Woodlands date, for the most part, to the late Pleistocene–early Holocene (12,500–8,000 yr B.P.), and belong to the Paleoindian cultural-historical period (for authoritative recent summaries of Eastern Woodlands prehistory, see Smith [1986] and Steponaitis [1986]). These people were foragers whose general lifeway probably closely resembled that of their not-so-distant ancestors, the first humans to enter the

Western Hemisphere from northeastern Asia via Beringia. The Paleoindians and their successors of the initial Early Archaic bore the brunt of coping with late glacial and postglacial conditions. By Middle Holocene time (8,000–5,000 yr B.P.), which is roughly coterminous with the Middle Archaic, climate, flora, and fauna in the mid- and southerly latitudes of eastern North America were similar to those of today, except for the effects of the Hypsithermal interval.

Except for dogs—which probably have considerable antiquity as they are very common in Late Archaic shell-mound cemeteries—there are no prehistoric animal domesticates in eastern North America. Hence, culturally induced disturbance of vegetation patterns in eastern North America is significantly different from that in the Holocene forests of northern Europe and the Mediterranean woodlands in the Near East, where domestic cattle, sheep, and goats as well as early cultivational practices profoundly affected the natural vegetation. Nevertheless, the record of *plant* cultivation and domestication in eastern North American is extensive.

Three relevant pieces of data have recently emerged from the Middle Archaic archaeobotanical record: the first evidence for mass-processing of hickory nuts and subsequent availability of hickory as a widespread staple food, together with acorn and other forest foods; the first pre-3,000 yr B.P. evidence for the presence of presumably cultivated *Cucurbita* gourd and bottle gourd (*Lagenaria siceraria*) on or around human occupation sites; and the first evidence for pre-3,000 yr B.P. indigenous cultigens.

Although charred hickory nutshell is present in Early Archaic deposits, it is not until the Middle Archaic that it is found in the appropriate quantity and condition (smashed into small fragments) to indicate systematic exploitation in a manner documented ethnohistorically for the southeastern United States. One obvious controlling factor is that hickories and other nut-bearing trees were not abundantly available until the Eastern Woodlands took on their Middle Holocene character. Once this happened, human populations apparently developed the necessary techniques fairly rapidly to exploit them to the fullest, especially the oily, palatable, and nutritious (but very difficult to extract) hickory nutmeats. Although there is some variation from species to species, virtually all hickories have thick and highly involuted shells, making removal of the nutmeat frustratingly tedious and time-consuming. However, Middle Archaic populations discovered that the meats can be separated quickly from the shells by smashing the nuts en masse and dropping them into hot water that could have been provided in hide-lined depressions using hot-rock methods, for example (Munson, 1986, 1988; Talalay *et al.*, 1984). The meats and the oil float while the dense shell fragments sink. Historically known tribes in the southeastern U.S. greatly prized the oil, or "nutbutter," that can be concentrated by longer-term boiling of crushed hickory nuts (Dye, 1980; Hudson, 1976), and their prehistoric ancestors probably did, too, at least after thermally resistant pottery was in use, beginning about 2,000 yr B.P.

Middle Archaic deposits at two sites in southern Illinois (Napoleon Hollow and Koster) have yielded charred fragments of *Cucurbita* rind dating 7,100 ± 300 yr B.P., 7,000 ± 250 yr B.P., 6,860 ± 80 yr B.P. (all ^{14}C accelerator mass spectrometer ["direct"] dates; Asch and Asch, 1985; Conard *et al.*, 1984), and a *Cucurbita* seed from the Anderson site in Tennessee dates to 6,990 ± 120 yr B.P. (AMS date; Gary Crites, personal communication to Richard A. Yarnell, November 1987). Bottle gourd (*Lagenaria siceraria*) rind fragments from the Windover site in Florida date to 7,290 ± 120 yr B.P. (AMS date; Newsom, 1988). Sumpweed (*Iva annua*) seeds, cultivar size, from Napoleon Hollow date to 4,500 ± 500 yr B.P. (AMS date) and 1,970 ± 90 yr B.P. (conventional ^{14}C date); and cultigen *Chenopodium* from Newt Kash and Cloudsplitter rockshelters in eastern Kentucky date 3,400 ± 150, and 3,450 ± 150 yr B.P., respectively (AMS dates; Smith and Cowan, 1987). In the succeeding millennium and a half (covering the Late Archaic and Early Woodland periods), several other plants were taken into cultivation (e.g., Fritz, 1986; Gremillion and Ison, 1989), so that by Middle Woodland times (ca. 2,000 yr B.P.), many human groups in the Midwest and Midsouth were horticulturists as well as foragers and harvesters. The most important plant domesticates seem to have been sumpweed, sunflower (*Helianthus annuus*), *Chenopodium*, maygrass (*Phalaris caroliniana*), little barley (*Hordeum pusillum*), knotweed (*Polygonum erectum*), and the two gourd species (*Cucurbita pepo* var. *ovifera* and *Lagenaria siceraria*). The gourds (used primarily as containers, rattles, net floats, etc., rather than as food) are often thought to have been first domesticated in Latin America, but they may actually have been indigenous to the Eastern Woodlands (Decker, 1988; Heiser, 1985; King, 1985; Smith, 1987a, 1989, in press).

In spite of the fact that a well-documented suite of cultivars is present at many Early and Middle Woodland sites, there was still heavy reliance on nuts (especially hickory and acorn) and on other forest foods—plant and animal—as well as a variety of aquatic plant and animal species. The emphasis varied a great deal from region to region, and from season to season. In the Salts Cave/Mammoth Cave area of west central Kentucky, for example, where paleofecal evidence is abundant, we know that the proportion of plant food from cultivated species was well over 50%, at least seasonally (Watson, ed., 1969, 1974; Yarnell, 1974). But up to, and at the time of historical contact (16th and 17th centuries), there was considerable dependence by most groups on wild foods.

It now appears that the New World's tropical cultigen par excellence, maize (*Zea mays*), was not present in the Eastern Woodlands until about 2,000 years ago (Chapman and Crites, 1987; Conard *et al.*, 1984; Ford, 1987; Wagner, 1987; Watson, 1988, 1989; Wymer, 1987; Yarnell, 1986). It was introduced from Mexico at that time in a high row-numbered form variously referred to as

Midwestern 12-row, Chapalote, North American Pop, or Tropical Flint, but it was not used intensively before about 1,000 yr B.P. (900 a.c.e.). From then until the arrival of the European explorers, maize agriculture together with the cultivation of other Latin American domesticates (squashes and pumpkins [*Cucurbita pepo*], beans [*Phaseolus vulgaris*], and several of the older indigenous crops [e.g., sunflower, *Chenopodium*]) were extremely important in the Mississippi River valley and most of its major tributaries. Although the evidence is far too meager to be definitive, it appears that a lower row-numbered maize variety (called Northern Flint, or, as found archaeologically, Eastern 8-row) able to flourish in latitudes far north of the Mexican Border was not developed until ca. 800 a.c.e. Once that was accomplished (presumably by the people who lived in those northerly regions), maize, together with beans, became central to the economy of burgeoning chiefdoms (the Fort Ancient peoples) in the Ohio River drainage from southeastern Indiana to West Virginia. Yet at this same time (the half-millennium or so before European contact), densely inhabited, Middle Mississippian town-and-temple complexes along the Mississippi River from its junction with the Missouri to its lower reaches in Mississippi and Alabama were supported by a more diversified agricultural system (several forms of Midwestern 12-row maize, and the older, indigenous starchy and oily-seeded plants, plus squashes, pumpkins, and gourds), in addition to wild terrestrial and aquatic plants and animals (Blake, 1986; Wagner, 1987; Watson, 1988, 1989). Outside the Middle Mississippian/Fort Ancient spheres, many groups subsisted by means of foraging and harvesting, combined with varying amounts of horticulture.

The Origins of Horticulture in the Eastern Woodlands: Explanations and Interpretations

Discussion of early horticulture in eastern North America has for many decades been in the shadow of a larger issue, that of relations with and dependence on Mesoamerica. Maize (*Zea mays*), squashes and pumpkins (*Cucurbita pepo*), bottle gourds (*Lagenaria siceraria*), and beans (*Phaseolus vulgaris*) are all warm-temperate and tropical domesticates with long histories in Mesoamerica and South America. The impressive social, political, and ideological achievements of the Mississippians have often been attributed, in part at least, to influence of various kinds (stimulus diffusion, trade, or even migration) from Mexico. And it has long been known that maize, an important Mississippian staple crop, is Meso- or South American in origin. Hence, until very recently many scholars tended to view the development of horticulture and agriculture north of the border as owing to stimulus—if not actual seeds—from Mexico. This attitude was reinforced some 12 years ago when *Cucurbita pepo* was identified in mid-5th millennium B.P. (3rd millennium b.c.e.) deposits in Kentucky and Missouri

(Chomko and Crawford, 1978; Crawford, 1982), and then in much earlier horizons in Illinois, AMS direct-dated to ca. 7,000 yr B.P. (5,000 b.c.e.) (Asch and Asch, 1985; Conard *et al.*, 1984), three millennia older than any known cultigen indigenous to the Eastern Woodlands. However, the issue is still being debated (Decker, 1988; Heiser, 1985; King, 1985; Smith, 1987a), and there is not a consensus at the moment about the precise identity of the oldest (pre-4,000 yr B.P./2,000 b.c.e.) *Cucurbita*. It has been suggested that these specimens are not the Mesoamerican *C. pepo*, but rather another *Cucurbita* (such as the buffalo gourd, *C. foetidissima,* or the Texas gourd, *C. texana*), which may not even have been cultivated, let alone domesticated; or they may be indigenous cultigens derived from a *C. texana* variety that was more widely distributed throughout the Midwest prehistorically than is now the case.

The report of bottle gourd at a prehistoric site in Florida, direct-dated to ca. 7,000 yr B.P. (Newsom, 1988) is highly relevant but too recent to assess adequately. The oldest well-established finds of *C. pepo* gourd and *L. siceraria* (bottle gourd) elsewhere are mid-5th millennium B.P. (3rd millennium b.c.e.) (Phillips Spring, Missouri—*C. pepo* and *L. siceraria* [Kay, 1988; Kay *et al.*, 1980]; Carlston Annis, Kentucky—*C. pepo* [Crawford, 1982]; Bacon Bend, Tennessee—*C. pepo* [Chapman and Shea, 1981]) and mid-3rd millennium B.P. (first millennium b.c.e.) (Salts Cave, Kentucky—*C. pepo* and *L. siceraria* [Watson, ed., 1969]). Eastern North American maize is not much earlier than 1,800 yr B.P. (200 a.c.e.) (Chapman and Crites, 1987; Ford, 1987), and beans come in much later, ca. 900 yr B.P. (1,000 a.c.e.) or after (Wagner, 1987). Thus, the first occurrences of these undeniably imported cultigens are strung throughout the archaeobotanical record, as it is presently understood. The oldest previously uncontested identification of *C. pepo* at about 4,500 yr B.P. (2,500 b.c.e.) in western Missouri is at the Phillips Spring site; this material is currently being reevaluated and is not much earlier than the appearance of domesticated sumpweed at about 4,000 yr B.P. (2,000 b.c.e.) Moreover, it is generally agreed that this early *C. pepo* variety was a gourdlike form closely related to (if not identical with) the contemporary variety *ovifera*. That is, it would have been primarily a container (though the seeds are edible, nutritious, and palatable, the flesh is not) rather than a food. Therefore, it is quite possible to mount a strong argument to the effect that the origin of plant cultivation in eastern North America was an indigenous and autonomous process, independent of events in and imports from south of the border. Such an argument has recently been presented by Smith (1987a, 1989 in press), who notes the points summarized above: ambiguous identity of the oldest cucurbit; similar antiquity for the oldest widely accepted *Cucurbita pepo* remains and the oldest indigenous cultigen; and differential functions of the cucurbits—containers—vs. the indigenous cultigen, food. Smith evokes a plausible sequence of processes and events resulting in the independent derivation of domesticated food plants from local wild species in the Eastern Woodlands.

Smith's 1987 formulation begins with the postulated effects on midwestern riverine systems of the Hypsithermal interval, which was coterminous with much of the Middle Archaic. During this interval, midlatitude river regimes are thought to have stabilized, resulting in the creation and long-term maintenance of meanders, oxbow lakes, backwater lagoons, shoals and bars, and other slack-water situations previously nonexistent or ephemeral. These locales, being optimal for readily accessible, abundant, and diverse aquatic resources (mussels, fish, waterfowl, small mammals, roots, rhizomes, greens, etc.) were host to semisedentary and wholly sedentary human groups, who also ranged into the wooded uplands in systematic, seasonal pursuit of deer, turkey, and other forest animals as well as nuts and other wild plant foods. The stable openings created by semipermanent and permanent human habitation in floodplain locales were colonized by a variety of pioneer plant species each spring, some of which, such as sumpweed and *Chenopodium,* have edible and nutritious seeds. With minimal encouragement, these species could have been and were domesticated. Thus, Smith's account relies on environmental change and an interplay (of the sort that is now called coevolutionary after Rindos, 1984) between foraging-harvesting human groups and certain plant populations.

In another recent discussion, Brown (1985) uses a similar framework for considering the appearance of sedentism in the long and rich archaeological record from west central Illinois where permanent structures are present by 7,000 yr B.P. (5,000 b.c.e.) (at the Koster site, with evidence for "prolonged periods of encampment" dating back to 8,500 yr B.P./6,500 b.c.e. there), and where large base camps containing multiple storage pits are a prominent part of the Late Archaic archaeological record. Although he notes the first appearance of cultivated plants at the same time as the permanent structures (he is referring to the *Cucurbita* fragments from Koster and Napoleon Hollow, which—as indicated above—are not now accepted by everyone as indubitable domesticates), he does not place much weight on the role of early horticulture in prehistoric midwestern sociocultural evolution. Rather, his concern is focused on the effects of sedentism per se. He states that sedentism did not result from population pressure, which is a minor factor in the Archaic Midwest, and that the appearance of stable, resource-rich locales in the floodplains—at least in west central Illinois—preceded total commitment to sedentism by many hundreds of years (on the order of two millennia). The shift from mobile foraging to sedentary harvesting was, thus, very slow with a long transitional period of semimobile existence intervening when specific habitats (lagoons and swamps in the floodplain and woodlands at the upper edges of the floodplain) were exploited for their faunal and floral resources. Early horticulture does not make much of an impact in this Illinois River valley sequence until the Middle Woodland period ca. 2,150 yr

B.P. (150 b.c.e.), when a cultigen complex somewhat like those of the Early Woodland in Kentucky and Tennessee (but differing in several details) appears rather suddenly and becomes ubiquitous (Asch and Asch, 1985; Smith, 1987b). Long before that, however, beginning in the Middle Archaic, there is long-distance trade (in copper, marine shell, chert, and minerals) and differential patterning in mortuary treatment (location of graves and—in the Late Archaic—presence or absence of elaborate grave goods; Smith, 1986; Steponaitis, 1986).

Brown interprets the data described sketchily above as reflecting a gradual shift from autonomous local groups relying on mobility for subsistence security to forging of intergroup ties (represented by the exotic trade goods) facilitating mutual aid between sedentary societies in times of local resource shortages. Thus, social and political institutions take the place of mobility as a means of insurance against short-term, local shortage of vital resources. The stage is then set for the development of middle-range, nonegalitarian societies in which individuals compete for control of the trade goods, trade routes, or other special knowledge and commodities necessary to the intergroup exchange systems. The more successful competitors become "Big Men" or chiefs, status begins to shift from achieved to ascribed, and the social organization changes from egalitarian to ranked.

Although she did not have the benefit of the latest information on plant use and early cultigen chronology, Bender (1985b; see also Bender, 1978, 1985a) takes a position similar to Brown's in her comparison of Mesolithic–Neolithic Brittany and the Adena–Hopewell record of the midwestern United States. She believes that Early and Middle Woodland societies were intensively gathering, hunting, and fishing wild species, with minimal reliance on small-scale horticulture toward the end of the Middle Woodland. It is now clear from rapidly accumulating archaeobotanical evidence (Asch and Asch, 1985; Cowan, 1985a, 1985b; Smith, 1987b, in press; Watson, 1985, 1988, 1989; Wymer, 1987, 1988; Yarnell, 1969, 1974) that many Early and Middle Woodland groups were effective food producers. Nevertheless, her basic position—that social imperatives rather than technological and economic ones are the crucial variables—is much the same as Brown's (see also Braun, 1977, 1988; Marquardt, 1985) and is worthy of careful consideration.

Another example of recent discussions about horticultural origins in eastern North America is provided by Prentice (1986), who focuses on cucurbit gardening. Beginning with the proposition that the first cultivars in the eastern United States are warm-temperate and tropical gourds (*Cucurbita pepo* var. *ovifera* and *Lagenaria siceraria*), he sets out to explain how and why this might have come about by focusing on individual actions, specifically the nature and behavior of shamanistic experts in Archaic North America. He suggests that the most likely innovators in Archaic societies were the headmen, the traders (exotic materials and items are a conspicuous part of many Late Archaic archaeological assemblages), and the shamans. Because the shamanistic use of gourds as rattles and medicine containers is so well documented ethnographically and ethnohistorically in North America, he thinks prehistoric shamans are the most likely people to have effected introduction and diffusion of these plants to northerly regions well beyond the locales where they were first domesticated. This is an ingenious—if perhaps overcomplex—formulation with respect to the early gourd cultivars, but fails to account adequately for the cultivation and selective breeding of the indigenous food plants such as sumpweed, *Chenopodium,* and sunflower.

Watson and Kennedy (1991) also discuss the origins of horticulture in the Eastern Woodlands from the perspective of individual actions, but they differ from the other authors discussed above in their explicit consideration of gender in the past and gender bias in the present. They note that the most ethnographically and ethnohistorically plausible, gender-sensitive formulation would recognize women as the cultivators and domesticators, yet the two latest discussions either deny women any role in the process (Prentice, 1986) or trivialize that role by emphasizing the ease and naturalness of the proposed coevolutionary trajectory to domestication (Smith, 1987a). Watson and Kennedy conclude that, though Prentice and Smith probably did not deliberately choose to assume androcentric positions with regard to this issue, their accounts nonetheless reveal some aspects of gender bias in archaeological interpretation that should be examined more closely.

Conclusions

With respect to the main theme of this chapter, I note first that all the field and analytical work on the research question discussed here (origins of food production) in both areas is—virtually by definition—interdisciplinary and paleoecological (hereafter IPE). Further, it is only very recently that non-IPE factors have been adduced in any significant way to aid in explaining and interpreting the data on the processes they represent.

With respect to the databases in the two world areas, evidence for the development of plant cultivation in eastern North America is in general more abundant and more detailed than that from western Asia, largely for logistical reasons. There are some solid bodies of information for the Near East (mostly from the Levant, but also from Çayönü in southeastern Turkey and from the Deh Luran and Hamrin regions in Iran and Iraq, respectively), but the data points are much more widely scattered and the information less richly detailed. For the Eastern Woodlands there are three regions with very long (i.e., diachronic) archaeobotanical records beginning before the appearance of cultigens and continuing to the end of prehistoric times: west central Illinois (Asch and Asch, 1985), the American Bottom in Illinois (Johannessen, 1984), and central and eastern Tennessee (Brewer, 1973; Chapman and Crites, 1987; Chapman and Shea, 1981; Crites, 1978). In addition, there are four other regions with extraordinarily detailed synchronic information: the central Ohio River valley (Wagner, 1987; Wymer, 1987), eastern Kentucky (Cowan, 1985b; Gremillion and Ison, 1989; Ison, 1986, Smith and Cowan, 1987), western Kentucky (Watson, ed., 1969, 1974; Watson and Yarnell, 1986; Yarnell, 1986), and the Ozarks (Fritz, 1986).

In the Near East comprehensive materials pertaining to the beginnings and earliest development of horticultural and herding societies have been obtained from Tell Aswad, Abu Hureyra, Mureybat, Çayönö, Ganj Dareh, and Ali Kosh. There is information about horticulture, agriculture, herding, and pastoralism from many more 8th millennium B.P. (6th millennium b.c.e.) and later sites in the Levant, Anatolia, and the Zagros, but very few provide the amount of detail that is available for later prehistory in the Eastern Woodland locales listed at the beginning of this section. A significant proportion of that detail derives from CRM projects, however, and perhaps when the numerous salvage excavations now in progress in Turkey, Syria, and Iraq are published, a much more robust corpus of data will be at hand for those regions as well.

With respect to interpretations and explanations for the origins of food production, the most detailed ones, those closest to the data, are—predictably—thoroughly IPE for both the Near East and eastern North America. Henry and Moore, however, refer to social factors above and beyond the brute issue of population increase, and Smith (1987b, 1989, in press) has recently taken up the issue of social evolution in the Woodland period of eastern North American prehistory. Furthermore, social imperatives dominate the discussion of several archaeologists studying eastern North America (Bender, 1985a, 1985b; Braun, 1977, 1988; Brown, 1985; Marquardt, 1985) and are beginning to influence the thinking of many others.

Redding (1988) provides a generalized formulation for the origin of food production that includes a number of topics relevant to both western Asia and eastern North America. His discussion is definitely IPE, but differs from others in that he gives specific attention to individual human decisions about limiting reproduction, to the relations among strategies and techniques alternative to food production, and to certain characteristics of the context (selective milieu, in his words) in which food production arises. Redding says that the crucial selective milieu is a cooccurrence of growing human populations that have utilized all tactics—except limiting reproduction—available to foragers and harvesters (these tactics are mobility; diversification of wild food sources, that is, adopting a broad-spectrum economy; and creation and use of storage facilities); that are at or near the carrying capacity of their environments; and that live in environments where fluctuations in food resources are unpredictable, frequent, and severe. Given such a milieu, food production will emerge because the only alternative is birth control, and birth control will not be applied systematically as a long-term strategy unless there is no other possible option: "An individual limiting his or her reproduction in response to some environmental stress in an area in which other individuals are reproducing, *successfully,* at a higher level because they have adopted a strategy or tactic that alleviates the stress, must either cease limiting his or her reproduction and adopt the alternative strategy, or be selected against" (Redding, 1988: 68). Several of the principles and implications of Redding's formulation contrast with those discussed for both western Asia and eastern North America, as indicated below.

Comparison of Interpretations: Western Asia and Eastern North America

If one compares current interpretive explanations for the two regions, one can make some interesting generalizations on two levels: first, about the substance or content of these formulations; second, about the nature and structure of the formulations themselves.

In both world areas, agricultural and pastoral economies developed from foraging economies via a lengthy (minimally, several millennia) period of foraging and harvesting subsistence systems. If viewed orthogenetically, the whole sequence for each area is therefore tripartite, with the first two stages much longer than the third. In eastern North America, the second, or foraging-and-harvesting stage, was so successful that it lasted into early historic (European contact) times. The late-prehistoric Mississippian economies of the central Mississippi River valley, for example, supported many thousands of people on a diverse subsistence base that featured large proportions of wild terrestrial and aquatic foods as well as a distinctive agricultural combination of some half dozen, ancient indigenous cultivars with several Latin American ones (i.e., sunflower, sumpweed, *Chenopodium,* little barley, knotweed, maygrass, pumpkins and squashes, gourds, maize, and beans). This combined economy (both wild and cultivated species) of the Middle Mississippian cultures contrasts with present understanding of the prehistoric Near East, where Moore (1985: 61) describes a shift at about 8,000 yr B.P. (6,000 b.c.e.). Before that time, food from plant and animal domesticates was often combined with significant amounts of wild plant and animal foods, but after 8,000 yr B.P. a fully agricultural-pastoral economy becomes widespread with primary reliance on domestic wheat, barley, legumes, sheep, goat, pig, and cattle, and corresponding neglect of wild plant and animal foods. In eastern North America, only the Fort Ancient Mississippians of the central Ohio River valley are known to display such a clear-cut and widespread shift from a varied subsistence base to a more restricted one dominated by a few domesticated plants. This shift occurs quite rapidly at about 1,000 yr B.P. (1,000 A.C.E.) and—though recently well-documented archaeobotanically (Wagner, 1987; Wymer, 1987)—is not yet well understood.

In both the Near East and eastern North America, a long period of sedentism apparently precedes the first food-producing economies (Moore [1985] disagrees with this [as apparently does Hole, 1984; see below], but—as noted above—he has now altered his opinion as a result of the latest information from Abu Hureyra [Moore, 1988]). The phenomena of sedentism and of population increase are critical discussion points for both regions. Henry (1983, 1989) views sedentism (and subsequent population increase) as a natural result of the emergence of optimal resource zones in the early Holocene period in the Levant. His formulation thus resembles some basic aspects of Braidwood's earlier interpretations (1954, 1975) for the Near East, and of Smith's (1987a) for eastern North America. Moore (1985) also thinks human population increase naturally resulted from early Holocene environmental amelioration followed by sedentism and agglomeration into villages as denser populations caused crowding relative to earlier conditions. Further population increase took place as a response to sedentary life, and the pressure on wild plants and animals caused subsistence intensification in the direction of domestication. In Moore's 1985 interpretation, sedentism and food production are essentially synchronous, but as just noted he no longer holds this opinion (Moore, 1988). Hole (1984: 50, 56–57) believes that sedentism and population increase result from stage two of the domestication processes he postulates (i.e., convergence and effective combination of agricultural and pastoral economies), and hence thoroughgoing or fully effective sedentism does not precede the creation of domestic species in his model. Redding (1988: 82) downplays sedentism, which he sees as simply a potential source of selective pressure on subsistence behavior because it permits closer spacing of births and hence population increase.

In eastern North America it is clear that in some places long-term, seasonal sedentism (i.e., long-term return to seasonal base camps from which task groups move out occasionally to smaller, more transitory, special-purpose camps in a pattern Henry calls radiation, vs. more mobile "circulation" systems) and even year-round sedentism precedes early horticulture (Brown, 1985; Braun, 1988; Smith, 1986, 1987a; Steponaitis, 1986). Furthermore, Smith (1986) for eastern North America and Henry (1983, 1985, 1989) for the Levant suggest that some form of social ranking (Big Men or chiefdomlike societies arose either just before (the Levant) or synchronously with (midwestern North America) the earliest food-producing economies, whereas Moore (1985) believes ranked societies did not appear in the Near East until well after the establishment of agriculture and pastoralism.

An important similarity between the two regions—because it is probably much more widely generalizable—is the diversity of subsistence systems at any one time in the periods under investigation. Although it is perfectly possible to construct orthogenetic developmental sequences from full-scale foraging to intensive food production on the basis of the present evidence in both areas, it is certainly the case that a cross section at any point in such a sequence would reveal a spectrum of foraging, foraging-and-harvesting, and—in the later part of the sequence—early food-producing communities flourishing in coexistence and interacting with each other. The Eastern Woodlands with their richer biotic diversity (especially of aquatic species) are perhaps more striking in this regard, but such a spectrum, or mosaic, must have been present in the Near East also, as indicated by the archaeobotanical information from the Levant. As more high-quality data are recovered from Near Eastern sites and subjected to increasingly sophisticated analyses (e.g., Hillman, 1988; Hillman *et al.*, 1986; van Zeist and Bakker-Heeres, 1982 [1985], 1984 [1986]), a finely divided spectrum of complex, local subsistence economies will surely be revealed, as they are being so revealed in eastern North America.

As to the nature and structure of explanatory interpretations in the two areas, it is clear that IPE formulations are dominant, that prime-mover (essentially monocausal) constructions have given way to multivariate ones, and that social imperatives are now being taken seriously for both eastern North America and the eastern Mediterranean. Explicit application of more overtly post-processualist approaches to the specific data considered here has so far been attempted in print by only one archaeologist (Prentice, 1986), but there will probably be others (e.g., Watson and Kennedy, 1991). All empirically based, future explanations of whatever category, however—processual, post-processual, or other—are, in a very real sense, simply epiphenomenal elaboration on the sturdy edifice of knowledge created by interdisciplinary, paleoecological research.

Acknowledgments

I am grateful to Linda Shane and the other symposium organizers for inviting me to contribute to the original session and the subsequent publication. The following people generously assisted me by providing key references, reprints, and/or access to data that are not yet published: Peter M.M.G. Akkermans, David Dye, David Harris, Gordon Hillman, Keith Kintigh, Lee Newsom, Andrew Moore, Bruce Smith, Maurits van Loon, Willem van Zeist, Gail Wagner, and Richard Yarnell. In addition, David Harris, Maurits van Loon, and Willem van Zeist kindly allowed me to use the libraries and other facilities of their respective institutions: the Department of Human Environment at the University of London's Institute of Archaeology, the West Asian Department of the

University of Amsterdam's Pre- and Protohistoric Institute, and the University of Gronigen's Biological-Archaeological Institute. Linda and Orrin Shane skillfully copyedited the original manuscript, and—together with two anonymous reviewers for the University of Minnesota Press—made several suggestions for its revision. I am thankful for their care and patience. The scholars named here are not responsible for the final product, but it was significantly enhanced by their assistance.

References

Asch, D. L., and Asch, N. B. (1985). Prehistoric plant cultivation in west-central Illinois. *In* "Prehistoric Food Production in North America" (R. Ford, Ed.), pp. 149–204. University of Michigan, Museum of Anthropology, Anthropological Papers 75, Ann Arbor.

Bareis, C., and Porter, J., Eds. (1984). "American Bottom Archaeology." University of Illinois Press, Urbana.

Bar-Yosef, O. (1980). Prehistory of the Levant. *Annual Review of Anthropology* **9**, 101–133.

Bender, B. (1978). Gatherer-hunter to farmer: a social perspective. *World Archaeology* **10**, 204–222.

——. (1985a). Emergent tribal formations in the American midcontinent. *American Antiquity* **50**, 52–62.

——. (1985b). Prehistoric developments in the American midcontinent and in Brittany, northwest France. *In* "Prehistoric Hunter-Gatherers: The Emergence of Cultural Complexity" (T. Price and J. Brown, Eds.), pp. 21–57. Academic Press, Orlando, FL.

Bense, J. A., Ed. (1983). "Archaeological Investigations in the Upper Tombigbee Valley, Mississippi: Phase I." University of West Florida, Office of Cultural and Archaeological Research, Reports of Investigations 3 (4 vols.), Pensacola.

Binford, L. R. (1962). Archaeology as anthropology. *American Antiquity* **28**, 217–225.

——. (1968). Post-Pleistocene adaptations. *In* "New Perspectives in Archeology" (S. Binford and L. Binford, Eds.), pp. 313–341. Aldine, Chicago.

Bintliff, J. L., and van Zeist, W., Eds. (1982). "Palaeoclimates, Paleo-Environments and Human Communities in the Eastern Mediterranean Region in Later Prehistory." British Archaeological Reports International Series 133, Oxford.

Blake, L. (1986). Corn and other plants from prehistory into history in the eastern United States. *In* "The Protohistoric Period in the Mid-South: 1500–1700" (D. Dye and R. Brister, Eds.), pp. 3–13. Proceedings of the 1983 Mid-South Archaeological Conference, Mississippi Department of Archives and History, Archaeological Report 18.

Braidwood, L. S., and Braidwood, R. J. (1982). "Prehistoric Village Archaeology in South-Eastern Turkey." British Archaeological Reports International Series 138, Oxford.

——. (1986). Prelude to the disappearance of village-farming communities in southwestern Asia. *In* "Ancient Anatolia: Aspects of Change and Cultural Development (Essays in Honor of Machteld J. Mellink)" (J. Vorys, E. Porada, B. Ridgway, and T. Stech, Eds.), pp. 3–11. University of Wisconsin Press, Madison.

Braidwood, L. S., Braidwood, R. J., Howe, B., Reed, C. A., and Watson, P. J., Eds. (1983). "Prehistoric Archaeology Along the Zagros Flanks." Oriental Institute, University of Chicago, Oriental Institute Publication 105.

Braidwood, R. J. (1954). "The Near East and the Foundations of Civilization." Condon Lectures, Oregon System of Higher Education, Eugene.

——. (1960). The agricultural revolution. *Scientific American* **203**, 130–148.

——. (1975). "Prehistoric Men." Scott, Foresman, Glenview, IL.

Braidwood, R. J., and Howe, B., Eds. (1960). "Prehistoric Investigations in Iraqi Kurdistan." Oriental Institute, University of Chicago, Studies in Ancient Oriental Civilization 31, Chicago.

Braidwood, R. J., Çambel, H., and Schirmer, W. (1981). Beginnings of village-farming communities in southeastern Turkey: Çayönü Tepesi, 1978 and 1979. *Journal of Field Archaeology* **8**, 249–258.

Braun, D. (1977). "Middle Woodland—(Early) Late Woodland Social Change in the Prehistoric Central Midwestern United States." Unpublished Ph.D. dissertation, University of Michigan. University Microfilms 77-26, 210, Ann Arbor.

——. (1988). Social and technological roots of "Late Woodland." *In* "Interpretations of Culture Change in the Eastern Woodlands during the Late Woodland Period" (R. Yerkes, Ed.), pp. 17–38. Occasional Papers in Anthropology 3, Ohio State University, Columbus.

Brewer, A. J. (1973). Analysis of floral remains from the Higgs site (40 LO 45). *In* "Excavation of the Higgs and Doughty Sites: I-75 Salvage Archaeology" (M. McCollough and C. Faulkner, Eds.), pp. 141–144. Tennessee Archaeological Society Miscellaneous Papers 12, Knoxville.

Brown, J. (1985). Long term trends to sedentism and the emergence of complexity in the American Midwest. *In* "Prehistoric Hunter-Gatherers: The Emergence of Cultural Complexity" (T. Price and J. Brown, Eds.), pp. 201–234. Academic Press, Orlando, FL.

Butzer, K. (1975). The "ecological" approach to prehistory: Are we really trying? *American Antiquity* **40**, 106–111.

——. (1978). Toward an integrated contextual approach in archaeology. *Journal of Archaeological Science* **5**, 191–193.

——. (1980). Context in archaeology: An alternative perspective. *Journal of Field Archaeology* **7**, 417–422.

Byers, D., Ed. (1967). "The Prehistory of the Tehuacan Valley, Volume I." University of Texas Press, Austin.

Cauvin, J. (1977). Les fouilles de Mureybet (1971–1974) et leur signification pour les origines de la sedentarisation au Proche-Orient. *Annual of the American Schools of Oriental Research* **44**, 19–48.

——. (1978). "Les premiers villages de Syrie-Palestine du IXème au VIIème millénaire avant J.C." Maison de l'Orient, Lyon.

Chapman, J. (1985). "Tellico: 12,000 Years of Native American History." University of Tennessee Press, Knoxville.

Chapman, J., and Crites, G. (1987). Evidence for early maize (*Zea mays*) from the Icehouse Bottom site, Tennessee. *American Antiquity* **52**, 352–354.

Chapman, J., and Shea, A.B. (1981). The archaeobotanical record: Early Archaic to contact in the lower Little Tennessee River Valley. *Tennessee Anthropologist* **6**, 64–84.

Chomko, S. A., and Crawford, G. W. (1978). Plant husbandry in prehistoric eastern North America: New evidence for its development. *American Antiquity* **43**, 405–408.

Clark, G. (1952). "Prehistoric Europe: The Economic Basis." Cambridge University Press, Cambridge, England.

Clarke, D. L. (1968). "Analytical Archaeology." Methuen, London.

Clutton-Brock, J. (1979). The mammalian remains from the Jericho tell. *Proceedings of the Prehistoric Society* **45**, 135–157.

Cohen, M. N. (1977). "The Food Crisis in Prehistory." Yale University Press, New Haven, CT.

Conard, N., Asch, D., Asch, N., Elmore, D., Gove, H., Rubin, M., Brown, J., Wiant, M., Farnsworth, K., and Cook, T. (1984). Accelerator radiocarbon dating of evidence of prehistoric horticulture in Illinois. *Nature* **308**, 443–446.

Conkey, M., and Spector, J. (1984). Archaeology and the study of gender. *In* "Advances in Archaeological Method and Theory" (M. Schiffer, Ed.), vol. 7, pp. 1–38. Academic Press, Orlando, FL.

Cowan, C. W. (1985a). Understanding the evolution of plant husbandry in eastern North America:

Lessons from botany, ethnography, and archaeology. *In* "Prehistoric Food Production in North America" (R. Ford, Ed.), pp. 205–244. University of Michigan, Museum of Anthropology, Anthropological Papers 75, Ann Arbor.

——. (1985b). "From foraging to incipient food-production: Subsistence change and continuity on the Cumberland Plateau of eastern Kentucky." Unpublished Ph.D. dissertation, University of Michigan, Ann Arbor.

Crawford, G. W. (1982). Late Archaic plant remains from west-central Kentucky: A summary. *Midcontinental Journal of Archaeology* **7**, 205–224.

Crites, G. (1978). "Paleoethnobotany of the Normandy Reservoir in the Upper Duck River Valley, Tennessee." Unpublished M.A. thesis, University of Tennessee, Knoxville.

Daniel, G. (1975). "150 Years of Archaeology." Duckworth, London.

Decker, D. (1988). Origin(s), evolution, and systematics of *Cucurbita pepo* (Cucurbitaceae). *Economic Botany* **42**, 4–15.

Delcourt, P. A., and Delcourt, H. R. (1979). Late Pleistocene and Holocene distributional history of the deciduous forest in the southeastern United States. *Veroffentlichungen des Geobotanischen Institutes der Eidgenoessische Technische Hochschule, Stiftung Ruebel (Zurich)* **68**, 79–107.

——. (1981). Vegetation maps for eastern North America. *In* "Geobotany II" (R. Romans, Ed.), pp. 123–165. Plenum, New York.

——. (1983). Late Quaternary vegetational dynamics and community stability reconsidered. *Quaternary Research* **19**, 265–271.

Dever, W. G. (1981). The impact of the "New Archaeology" on Syro-Palestinian archaeology. *Bulletin of the American Schools of Oriental Research* **242**, 15–29.

Ducos, P., and Helmer, D. (1981). Le point actuel sur l'apparition de la domestication dans la Levant. *In* "Prehistoire du Levant" (J. Cauvin, Ed.), pp. 523–528. Actes du Colloque International 598, Editions CNRS, Paris.

Dunnell, R. C. (1979). Trends in current Americanist archaeology. *American Journal of Archaeology* **83**, 437–449.

——. (1980). Americanist archaeology: The 1979 contribution. *American Journal of Archaeology* **84**, 463–478.

——. (1981). Americanist archaeology: The 1980 literature. *American Journal of Archaeology* **85**, 429–445.

——. (1982). Americanist archaeological literature: 1981. *American Journal of Archaeology* **86**: 509–529.

——. (1983). A review of the Americanist literature for 1982. *American Journal of Archaeology* **87**, 521–544.

——. (1984). The Americanist literature for 1983: A year of contrasts and challenges. *American Journal of Archaeology* **88**, 489–513.

——. (1985). Americanist archaeology in 1984. *American Journal of Archaeology* **89**, 585–611.

Dye, D. (1980). "Primary forest efficiency in the western middle Tennessee Valley." Unpublished Ph.D. dissertation, Washington University, St. Louis.

Dyson, S. L. (1981). A classical archaeologist's response to the "new archaeology." *Bulletin of the American Schools of Oriental Research* **242**, 7–13.

Evins, M. A. (1983). The fauna from Shanidar Cave: Moustarian wild goat exploitation in northeastern Iraq. *Paleorient* **8**, 37–58.

Faulkner, C. H., and McCullough, M. C., Eds. (1973). "Introductory Report of the Normandy Reservoir Salvage Project: Environmental Setting, Typology, and Survey." University of Tennessee Department of Anthropology, Reports of Investigations 11, Knoxville.

——. (1982). "Eighth Report of the Normandy Archaeological Project." University of Tennessee

Department of Anthropology, Reports of Investigations 33, Tennessee Valley Authority Publications in Anthropology 30, Knoxville.
Ford, R. I. (1987). Dating early maize in the eastern United States. Paper read at the 10th annual conference of the Society of Ethnobiology, Gainesville, FL.
Fritz, G. (1986). "Prehistoric Ozark Agriculture; the University of Arkansas Rockshelter Collections." Unpublished Ph.D. dissertation, University of North Carolina–Chapel Hill.
Gero, J. (1985). Socio-politics and the woman-at-home ideology. *American Antiquity* **50,** 342–350.
Gero, J., and Conkey, M., Eds. (1991). "Engendering Archaeology: Women and Prehistory." Oxford, Basil Blackwell Ltd.
Gero, J., Lacy, D., and Blakey, M., Eds. (1983). "The Socio-Politics of Archaeology." Anthropological Research Report, University of Massachusetts, Amherst.
Grayson, D. (1983). "The Establishment of Human Antiquity." Academic Press, New York.
Green, J., Ed. (1979). "Zuni: Selected Writings of Frank Hamilton Cushing." University of Nebraska Press, Lincoln and London.
Gremillion, K. J., and Ison, C. R. (1989). Terminal Archaic and Early Woodland plant utilization along the Cumberland Plateau. Paper presented at the 54th annual meeting of the Society for American Archaeology, April 5–9, Atlanta.
Hastorf, C., and Popper, V., Eds. (1989). "Current Paleoethnobotany: Analytical Methods and Cultural Interpretations of Archaeological Plant Remains." University of Chicago Press, Chicago.
Heiser, R. (1985). Some botanical considerations of the early domesticated plants north of Mexico. *In* "Prehistoric Food Production in North America" (R. Ford, Ed.), pp. 57–72. University of Michigan, Museum of Anthropology, Anthropological Papers 75, Ann Arbor.
Heizer, R. F. (1962). "Man's Discovery of his Past; Literary Landmarks in Archaeology." Prentice-Hall, Englewood Cliffs, NJ.
Helbaek, H. (1969). Plant collecting, dry-farming, and irrigation agriculture in prehistoric Deh Luran. *In* "Prehistory and Human Ecology of the Deh Luran Plain" (F. Hole, K. Flannery, and J. Neely, Eds.), pp. 383–426. University of Michigan Museum of Anthropology Memoirs 1.
Henry, D. O. (1983). Adaptive evolution within the Epipaleolithic of the Near East. *In* "Advances in World Archaeology" (F. Wendorf and A. Close, Eds.), vol. 2, pp. 99–160. Academic Press, London.
——. (1985). Preagricultural sedentism: The Natufian example. *In* "Prehistoric Hunter-Gatherers: The Emergence of Cultural Complexity" (T. Price and J. Brown, Eds.), pp. 365–384. Academic Press, Orlando, FL.
——. (1989). "From Foraging to Agriculture: The Levant at the end of the Oce Age." University of Pennsylvania Press, Philadelphia.
Hesse, B. (1979). Rodent remains and sedentism in the Neolithic: Evidence from Tepe Ganj Dareh, western Iran. *Journal of Mammalogy* **60,** 856–857.
——. (1982). Slaughter patterns and domestication: The beginnings of pastoralism in western Iran. *Man* **17,** 403–417.
Hillman, G. C. (1975). The plant remains from Tell Abu Hureyra: A preliminary report. Appendix A in A. Moore, The excavation of Tell Abu Hureyra in Syria: A preliminary report. *Proceedings of the Prehistoric Society* **41,** 70–73.
——. (1988). Plant food economy of the two settlements at Abu Hureyra: Dietary diversity, seasonality, and advent of agriculture. Paper presented at the 53rd annual meeting of the Society for American Archaeology, April 27-May 1, Phoenix.
Hillman, G. C., Colledge, S. M., and Harris, D. R. (1986). Plant-food economy during the Epi-Paleolithic period at Tell Abu Hureyra, Syria: Dietary diversity, seasonality, and modes of exploi tation. Paper presented at the World Archaeological Congress, September, Southampton, England.
Hodder, I. (1986). "Reading the Past." Cambridge University Press, Cambridge.

Hole, F. (1984). A reassessment of the Neolithic revolution. *Paleorient* **10**, 49–60.

Hole, F., Flannery, K., and Neely, J., Eds. (1969). "Prehistory and Human Ecology of the Deh Luran Plain." University of Michigan Museum of Anthropology, Memoir 1, Ann Arbor.

Hopf, M. (1969). Plant remains and early farming in Jericho. *In* "The Domestication and Exploitation of Plants and Animals" (P. Ucko and G. Dimbley, Eds.), pp. 355–359. Duckworth, London.

Hudson, C. (1976). "The Southeastern Indians." University of Tennessee Press, Knoxville.

Ison, C. (1986). Recent excavations at Cold Oak Shelter, Daniel Boone National Forest, Kentucky. Paper presented at the Kentucky Heritage Council annual conference, Louisville.

Jacobson, G. L., Jr., Webb, T., III, and Grimm, E.C. (1987). Patterns and rates of vegetation change during the deglaciation of eastern North America. *In* "North American and Adjacent Oceans During the Last Deglaciation: The Geology of North America," v. K-3 (W. Ruddiman and H. E. Wright, Jr., Eds.), pp. 277–288. Geological Society of America, Boulder, CO.

Johannessen, S. (1984). Paleoethnobotany. *In* "American Bottom Archaeology" (C. Bareis and J. Porter, Eds.), pp. 197–214. University of Illinois Press, Urbana.

Kay, M. (1987). Phillips Spring: A synopsis of Sedalia Phase settlement and subsistence. In "Foraging, Collecting, and Harvesting: Archaic Period Subsistence and Settlement in the Eastern Woodlands" (S. Neusius, Ed.), pp. 275–288. Southern Illinois University-Carbondale Center for Archaeological Investigations, Occasional Paper 6.

Kay, M., King, F., and Robinson, C. (1980). Cucurbits from Philips Spring: New evidence and interpretations. *American Antiquity* **45**, 806–822.

Kennedy, M. C. (1979). "Status, role, and gender: Preconceptions in archaeology." Honors Seminar thesis on file, Department of Anthropology, University of Minnesota, Minneapolis.

Kidder, A. V. (1924). "An Introduction to the Study of Southwestern Archaeology with a Preliminary Account of the Excavations at Pecos." Reprinting of 1962 with Introduction by Irving Rouse, Yale University Press, New Haven, CT.

——. (1937). The development of Maya research. Proceedings of the Second General Assembly, Pan American Institute of Geography and History, 1935: 218–235. U.S. Department of State Conference Series, no. 28.

King, F. (1985). Early cultivated cucurbits in eastern North America. In "Prehistoric Food Production in North America" (R. Ford, Ed.), pp. 73–97. University of Michigan, Museum of Anthropology, Anthropological Papers 75, Ann Arbor.

King, J. E. (1981). Late Quaternary vegetational history of Illinois. *Ecological Monographs* **5**: 43–62.

King, T., Hickman, P. P., and Berg, G. C. (1977). "Cultural Resource Management: An Anthropological Approach." Academic Press, New York.

Legge, A. J. (1988). Steppe hunting and the origins of animal domestication at Abu Hureyra. Paper presented at the 53rd annual meeting of the Society for American Archaeology, April 27-May 1, Phoenix.

Leone, M. P. (1978). Archaeology as the science of technology: Mormon town plans and fences. *In* "Research and Theory in Current Archaeology" (C. Redman, Ed.), pp. 125–150. Wiley, New York.

——. (1982). Some opinions about recovering mind. *American Antiquity* **47**, 742–760.

——. (1986). Symbolic, structural, and critical archaeology. In "American Archaeology, Past and Future: A Celebration of the Society for American Archaeology 1935–1985" (D. Meltzer, D. Fowler, J. Sabloff, Eds.), pp. 415–438. Smithsonian Institution Press, Washington and London.

Leroi-Gourhan, A. (1974). Etudes palynologiques des derniers 11.000 anos en Syrie semi-désertique. *Paléorient* **2**, 443–451.

Lyell, C. (1830–1833). "Principles of Geology." John Murray, London.

MacNeish, R. S. (1967). An interdisciplinary approach to an archaeological problem. *In* "The Prehistory of the Tehuacan Valley, Volume I" (D. Byers, Ed.), pp. 14–24. University of Texas Press, Austin.

Marquardt, W. H. (1985). Complexity and scale in the study of fisher-gatherer-hunters: An example from the eastern United States. *In* "Prehistoric Hunter-Gatherers: The Emergence of Cultural Complexity" (T. Price and J. Brown, Eds.), pp. 59–98. Academic Press, Orlando, FL.

McDonald, W. A., and G. Rapp, Eds. (1972). "The Minnesota Messenia Expedition." University of Minnesota Press, Minneapolis.

McGimsey, C. R. III (1972). "Public Archaeology." Seminar Press, New York.

Mellaart, J. (1975). "The Neolithic of the Near East." Thames and Hudson, London.

Miller, N. (in press). Origins of plant cultivation in the Near East. *In* "Origins of Agriculture in World Perspective" (C. Cowan and P. Watson, Eds.). Smithsonian Institution Press, Washington, D.C.

Molleson, T. I. (1988). The people of Abu Hureyra. Paper presented at the 53rd annual meeting of the Society for American Archaeology, April 27-May 1, Phoenix.

Moore, A. M. T. (1975). The excavation of Tell Abu Hureyra in Syria: A preliminary report. *Proceedings of the Prehistoric Society* **41**, 50–77.

——. (1985). The development of Neolithic societies in the Near East. *In* "Advances in World Archaeology" (F. Wendorf and A. Close, Eds.), vol. 4, pp. 1–69. Academic Press, London.

——. (1988). The excavation of Abu Hureyra. Paper presented at the 53rd annual meeting of the Society for American Archaeology, April 27-May 1, Phoenix.

Munson, P. J. (1986). Hickory silvaculture: A subsistence revolution in the prehistory of eastern North America. Paper presented at the conference "Emergent Horticultural Economies of the Eastern Woodlands," organized by W. F. Keegan, April 1986, Southern Illinois University, Carbondale.

——. (1988). Late Woodland settlement and subsistence in temporal perspective. *In* "Interpretations of Culture Change in the Eastern Woodlands during the Late Woodland Period" (R. Yerkes, Ed.), pp. 7–16. Occasional Papers in Anthropology 3, Ohio State University, Columbus.

Newsom, L. (1988). Paleoethnobotanical remains from a waterlogged Archaic period site in Florida. Paper presented at the 53rd annual meeting of the Society for American Archaeology, April 27-May 1, Phoenix.

Niklewski, J., and van Zeist, W. (1970). A Late Quaternary pollen diagram from northwestern Syria. *Acta Botanica Neerlandica* **19**, 737–754.

Noy, T., Legge, A. J., and Higgs, E. S. (1973). Recent excavations at Nahal Oren, Israel. *Proceedings of the Prehistoric Society* **39**, 75–99.

Olsen, S. (1988). The bone artifacts from Abu Hureyra. Paper represented at the 53rd annual meeting of the Society for American Archaeology, April 27-May 1, Phoenix.

Olszewski, D. (1988). Stone tool use at Abu Hureyra 1. Paper presented at the 53rd annual meeting of the Society for American Archaeology, April 27-May 1, Phoenix.

Perkins, D. (1964). Prehistoric fauna from Shanidar, Iraq. *Science* **144**, 1565–1566.

Prentice, G. (1986). Origins of plant domestication in the eastern United States: Promoting the individual in archaeological theory. *Southeastern Archaeology* **5**, 103–119.

Pumpelly, R., Ed. (1908). "Explorations in Turkestan. Expedition of 1904. Prehistoric Civilizations of Anau; Origins, Growth, and Influence of Environment." Vols. 1 and 2. Carnegie Institution of Washington, Washington, D.C.

Redding, R. W. (1988). A general explanation of subsistence change: From hunting and gathering to food production. *Journal of Anthropological Archaeology* **7**, 56–97.

Redman, C. L. (1978). "The Rise of Civilization." W. H. Freeman, San Francisco.

Rindos, D. (1984). "The Origins of Agriculture." Academic Press, New York.

Rowley-Conwy, P. (1988). Animal bones from Abu Hureyra: Methods of study and their results. Paper presented at the 53rd annual meeting of the Society for American Archaeology, April 27-May 1, Phoenix.

Ruddiman, W. F., and Wright, H. E., Jr., Eds. (1987). "North America and Adjacent Oceans During the Last Deglaciation: The Geology of North America," v. K-3. Geological Society of America, Boulder, CO.

Schiffer, M., and Gumerman, G. (Eds.) (1977). "Conservation Archaeology: A Guide for Cultural Resource Management Studies." Academic Press, New York.

Schoenwetter, J. A. (1981). Prologue to a contextual archaeology. *Journal of Archaeological Science* **8**, 367–379.

Smith, B. (1986). The archaeology of the southeastern United States: From Dalton to de Soto, 10,500–500 B.P. *In* "Advances in World Archaeology" (F. Wendorf and A. Close, Eds.), vol. 5, pp. 1–92. Academic Press, London.

——. (1987a). The independent domestication of indigenous seed-bearing plants in eastern North America. *In* "Emergent Horticultural Economies of the Eastern Woodlands" (W. Keegan, Ed.), pp. 3–47. Southern Illinois University–Carbondale Center for Archaeological Investigations, Occasional Paper 7.

——. (1987b). Hopewellian farmers of eastern North America. Paper presented at the 11th International Congress of Prehistoric and Protohistoric Sciences, August 31–September 5, Mainz, Germany.

——. (1989). Origins of agriculture in eastern North America. *Science* **246**, 1566–1571.

——. (in press). Prehistoric plant husbandry in eastern North America. *In* "Origins of Agriculture in World Perspective" (C. Cowan and P. Watson, Eds.). Smithsonian Institution Press, Washington, D.C.

Smith, B., and Cowan, C.W. (1987). The age of domesticated *Chenopodium* in prehistoric North America: New accelerator dates from eastern Kentucky. *American Antiquity* **52**, 355–357.

Smith, P. E. L. (1978). An interim report on Ganj Dareh Tepe, Iran. *American Journal of Archaeology* **82**, 538–540.

Solecki, R. (1981). An early village site at Zawi Chemi Shanidar. *Bibliotheca Mesopotamica* **82**, 538–540.

Steponaitis, V. (1986). Prehistoric archaeology in the southeastern United States, 1970–1985. *Annual Review of Anthropology* **15**, 363–404.

Struever, S. (1968). Flotation techniques for the recovery of small-scale archaeological remains. *American Antiquity* **33**, 353–362.

Talalay, L., Keller, L. D., and Munson, P. J. (1984). Hickory nuts, walnuts, butternuts, and hazelnuts: Observations and experiments relevant to their aboriginal exploitation in eastern North America. *In* "Experiments and Observations on Aboriginal Plant Food Utilization in Eastern North America" (P. Munson, Ed.), pp. 338–359. Indiana Historical Society Prehistory Research Series 6 (2).

Taylor, W. W. (1948). "A Study of Archaeology." American Anthropological Association Memoir 69. Southern Illinois University Press, Carbondale (1967).

Turnbull, P., and Reed, C.A. (1974). The fauna from the terminal Pleistocene of Palegawra Cave: A Zarzian occupation site in northeast Iraq. *Fieldiana: Anthropology* **63**, 81–146.

van Loon, M. (1968). The Oriental Institute excavations at Mureybit, Syria: Preliminary report on the 1965 campaign. Part I: Architecture and general finds. *Journal of Near Eastern Studies* **27**, 265–282.

van Zeist, W. (1967). Late Quaternary vegetation history of western Iran. *Review of Paleobotany and Palynology* **2**, 301–311.

——. (1972). Palaeobotanical results of the 1970 season at Çayönü, Turkey. *Helinium* **12**, 3–19.

van Zeist, W., and Bakker-Heeres, J. A. H. (1979). Some economic and ecological aspects of the plant husbandry of Tell Aswad. *Paleorient* **5**, 161–169.

——. (1982 [1985]). Archaeobotanical studies in the Levant 1. Neolithic sites in the Damascus Basin: Aswad, Ghoraife', Ramad. *Palaeohistoria* **24**, 165–256.

——. (1984 [1986]). Archaeobotanical studies in the Levant 3. Late-Paleolithic Mureybit. *Palaeohistoria* **26**, 171–199.

——. (1985 [1988]). Archaeobotanical studies in the Levant 4. Bronze Age sites on the North Syrian Euphrates. *Palaeohistoria* **27**, 247–316.

van Zeist, W., and Bottema, S. (1982). Vegetational history of the eastern Mediterranean and the Near East during the last 20,000 years. *In* "Palaeoclimates, Palaeoenvironments and Human Communities in the Eastern Mediterranean Region in Later Prehistory" (J. Bintliff and W. van Zeist, Eds.), pp. 277–321. British Archaeological Reports International Series 133, Oxford.

van Zeist, W., and Casparie, W. (1968). Wild einkorn wheat and barley from Tell Mureybit in northern Syria. *Acta Botanica Neerlandica* **17**, 44–53.

van Zeist, W., Smith, P. E. L., Palfenier-Vegter, R. M., Suwijn, M., and Casparie, W. (1986). An archaeobotanical study of Ganj Dareh Tepe, Iran. *Palaeohistoria* **26**, 201–224.

van Zeist, W., and Woldring, H. (1980). Holocene vegetation and climate of northwestern Syria. *Palaeohistoria* **22**, 111–125.

van Zeist, W., and Wright, H. E., Jr. (1963). Preliminary pollen studies at Lake Zeribar, Zagros Mountains, southwestern Iran. *Science* **140**, 65–67.

Wagner, G. E. (1982). Testing flotation recovery rates. *American Antiquity* **47**, 127–132.

——. (1987). "Uses of plants by the Fort Ancient Indians." Unpublished Ph.D. dissertation, Washington University, St. Louis.

——. (1989). Comparability among recovery techniques. *In* "Current Paleoethnobotany: Analytical Methods and Cultural Interpretations of Archaeological Plant Remains" (C. Hastorf and V. Popper, Eds.), pp. 17–35. University of Chicago Press, Chicago.

Watson, P. J. (1976). In pursuit of prehistoric subsistence: A comparative analysis of some contemporary flotation techniques. *Midcontinental Journal of Archaeology* **1**, 77–100.

——. (1985). The impact of early horticulture in the upland drainages of the Midwest and Midsouth. *In* "Prehistoric Food Production in North America" (R. Ford, Ed.), pp. 73–98. University of Michigan, Museum of Anthropology, Anthropological Papers 75, Ann Arbor.

——. (1986). Archaeological interpretation: 1985. *In* "American Archaeology: Past and Future. A Celebration of the Society for American Archaeology, 1935–1985" (D. Meltzer, D. Fowler, and J. Sabloff, Eds.), pp. 439–457. Smithsonian Institution Press, Washington, D.C.

——. (1988). Prehistoric gardening and agriculture in the Midwest and Midsouth. *In* "Interpretations of Culture Change in the Eastern Woodlands during the Late Woodland Period" (R. Yerkes, Ed.), pp. 38–66. Occasional Papers in Anthropology 3, Ohio State University, Columbus.

——. (1989). Early plant cultivation in the Eastern Woodlands of North America. *In* "Foraging and Farming: the Evolution of Plant Exploitation" (D. Harriss and G. Hillman, Eds.) pp. 555–571. Unwin Hyman, London.

——. (1990). The razor's edge: symbolic-structuralist archeology and the expansion of archeological inference. *American Anthropologist* **92**, 613–629.

Watson, P. J., Ed. (1969). "The Prehistory of Salts Cave, Kentucky." Illinois State Museum Reports of Investigations 16, Springfield.

——. (1974). "Archaeology of the Mammoth Cave Area." Academic Press, New York.

Watson, P. J., and Kennedy, M. C. (1991). The development of horticulture in the Eastern Woodlands: Women's role. *In* "Engendering Archaeology: Women and Prehistory" (J. Gero and M. Conkey, Eds.). Basil Blackwell, Oxford.

Watson, P. J., LeBlanc, S. A., and Redman, C. L. (1971). "Explanation in Archaeology: An Explicitly Scientific Approach." Columbia University Press, New York.

——. (1984). "Archaeological Explanation: The Scientific Method in Archaeology." Columbia University Press, New York.

Watson, P. J., and Yarnell, R. A. (1986). Lost John's last meal. *Missouri Archaeologist* **47**, 241–255.

Watts, W. A. (1980). The Late Quaternary vegetation history of the southeastern United States. *Annual Review of Ecology and Systematics* **11**, 387–409.

——. (1983). Vegetational history of the eastern United States 25,000 to 10,000 years ago. *In* "Late Quaternary Environments of the United States" (H. E. Wright and S. Porter, Eds.), vol. 1, pp. 294–310. University of Minnesota Press, Minneapolis.

Webb, T., III., Cushing, E. J., and Wright, H. E., Jr. (1983). Holocene changes in the vegetation of the Midwest. *In* "Late Quaternary Environments of the United States" (H. E. Wright and S. Porter, Eds.), vol. 2, pp. 142–165. University of Minnesota Press, Minneapolis.

Whitehead, D. R. (1973). Late-Wisconsin vegetational changes in unglaciated eastern North America. *Quaternary Research* **3**, 621–631.

Willey, G. R., and Phillips, P. (1958). "Method and Theory in American Archaeology." University of Chicago Press, Chicago.

Wright, G., and Miller, S. J. (1976). Prehistoric hunting of New World wild sheep: Implications for the study of sheep domestication. *In* "Cultural Change and Continuity" (C. Cleland, Ed.), pp. 293–312. Academic Press, New York.

Wright, H. E., Jr. (1976a). The dynamic nature of Holocene vegetation: A problem in paleoclimatology, biogeography, and stratigraphic nomenclature. *Quaternary Research* **6**, 581–596.

——. (1976b). Environmental setting for plant domestication in the Near East. *Science* **194**, 385–389.

——. (1977). Environmental change and the origin of agriculture in the Old and New Worlds. *In* "The Origins of Agriculture" (C. Reed, Ed.), pp. 281–318. Mouton, The Hague.

——. (1981). Vegetation east of the Rocky Mountains 18,000 years ago. *Quaternary Research* **15**, 113–125.

——. (1983). Climatic change in the Zagros Mountains—revisited. *In* "Prehistoric Archaeology Along the Zagros Flanks" (L. Braidwood, R. Braidwood, B. Howe, C. Reed, and P. Watson, Eds.), pp. 505–509. Oriental Institute Publication 105, Oriental Institute, University of Chicago.

Wright, H. E., Jr., Ed. (1983). "Late Quaternary Environments of the United States." Vols. 1 and 2. University of Minnesota Press, Minneapolis.

Wymer, D. A. (1987). The Middle Woodland-Late Woodland interface in central Ohio: Subsistence continuity amid cultural change. *In* "Emergent Horticultural Economies of the Eastern Woodlands" (W. Keegan, Ed.), pp. 201–216. Southern Illinois University–Carbondale, Center for Archaeological Investigations Occasional Paper 7.

——. (1988). Cultural change and subsistence: Middle Woodland-Late Woodland transition in the Midwest. Paper presented at the 53rd annual meeting of the Society for American Archaeology, April 27-May 1, Phoenix.

Yarnell, R. A. (1969). Contents of human paleofeces. *In* "The Prehistory of Salts Cave, Kentucky" (P. J. Watson, Ed.), pp. 41–54. Illinois State Museum Reports of Investigations 16, Springfield.

——. (1974). Plant food and cultivation of the Salts Cavers. *In* "Archaeology of the Mammoth Cave Area" (P. J. Watson, Ed.), pp. 113–122. Academic Press, New York.

——. (1983). Prehistory of plant foods and husbandry in North America. Paper presented at the 48th annual meeting of the Society for American Archaeology. Pittsburgh, April 27–30.

——. (1986). A survey of prehistoric crop plants in eastern North America. *The Missouri Archaeologist* **47**, 47–59.

2
The Last Interglacial / Glacial Cycle in Northern Europe

Jan Mangerud

Glacial Geology

The last interglacial/glacial cycle is a key period for understanding the earth's response to the orbital (Milankovitch) and other forcing of the climate, including the complex interplay between the atmosphere, ocean, glaciers, lithosphere, and vegetation. Because it is last, the cycle will always be the one that is known in most detail, and in fact the only one for which we can have any hope to work out a detailed glaciation history.

The scientific literature on the last interglacial and glacial in Europe is huge and can intimidate students and scientists not familiar with the area. One problem is that the results are published in several languages; another that publications occur in a large number of journals in geology, geography, botany, and so on. Still, the major problems are the volume of information and the difficulty of relating earlier observations and interpretations to new ideas and models. With this in mind, I have two major aims in this chapter.

1. To provide a review of the present knowledge and opinions about the last interglacial/glacial cycle in northern Europe, with emphasis on stratigraphy, general geological development, and correlation with the oceanographic record. This is a state-of-the-art review for those interested in what we really know about that cycle in probably the best-investigated area on the earth's continents.

2. To provide an introduction and a stratigraphic framework for those who want to penetrate deeper into the specialized literature, for example, on glacial development, vegetational history, paleoclimate, or more detailed regional studies.

The chapter is organized in three major parts. First, I examine the last interglacial and show it is the Eemian. What do we mean by the "last interglacial"? Can we identify and correlate sediment sequences that were laid down during that interglacial? How does the last interglacial in northern Europe correlate with the deep-sea oxygen-isotope stratigraphy (hereafter referred to as isotope stratigraphy), and thus with the stratigraphy over the rest of the world?

Second is a similar exercise with the last glacial stage, the Weichselian, in continental northwestern Europe. Here I concentrate on the area outside the glaciation limits to avoid the problems caused by glacial erosion. This stratigraphy is, to a large extent, the yardstick to which the history of the glaciated areas must be correlated.

Finally I move northward to the area glaciated by the Scandinavian ice sheet (Fig. 1), where the stratigraphy is much more fragmentary. Using a slightly different approach, I first describe and reexamine in some detail three key localities along the west coast of Norway that demonstrate how the history may be worked out in a glaciated area. Then I examine the evidence that core areas of the Scandinavian ice sheet were free of ice during periods of the Weichselian. A glaciation curve for both the western and eastern flank of the ice sheet is presented, though not examined in detail.

The Last Interglacial

WHAT IS AN INTERGLACIAL?

Most of us agree that we are living in an interglacial period. Yet there are two large ice sheets on the earth (Greenland and Antarctica) and some 1,750 glaciers in the Norwegian mountains alone (Liestøl, 1962). On the other hand, during the period traditionally called the last glaciation, Scandinavia was nearly as ice-free as today during one or two time intervals that we call interstadials and not

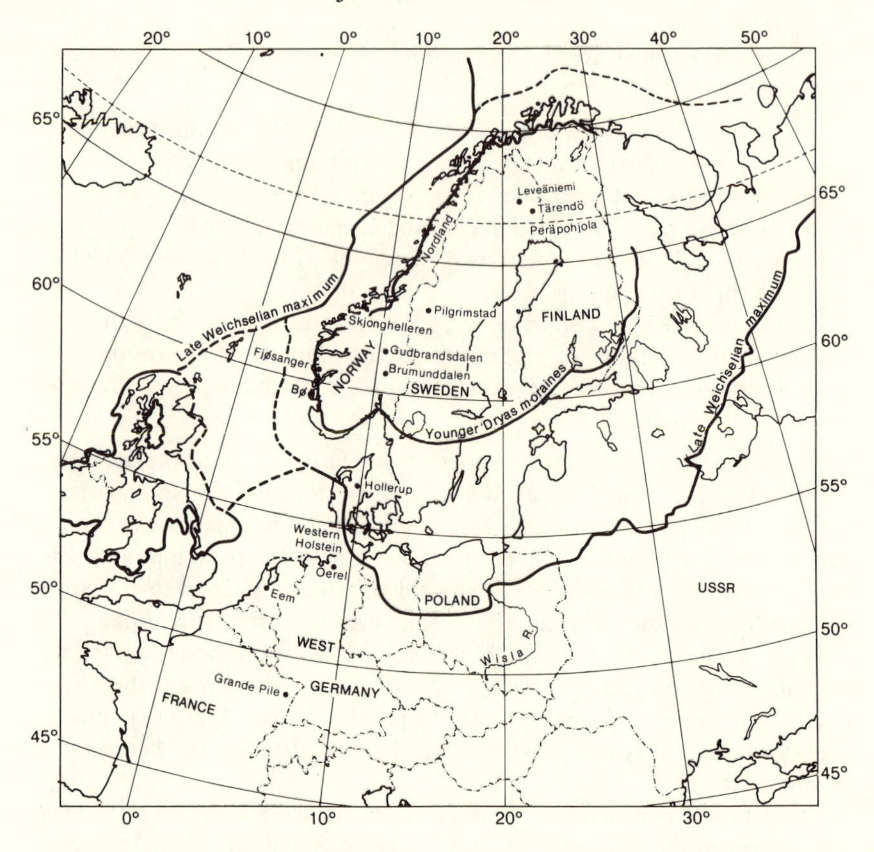

Fig. 1. Map of Europe showing the ice fronts for the Weichselian maximum and for the Younger Dryas for the Scandinavian and British ice sheets. The map includes localities mentioned in the text.

interglacials. Thus the definition of glacials and interglacials is not obvious (Suggate, 1974) and may in fact be meaningless further back in the Quaternary, when the amplitudes of climatic change were different.

The tradition in northern Europe is to relate interglacials to the climate and environments of the present, or rather of the Holocene. We label a period an interglacial if the *climate and duration* of the period allowed forest successions to proceed to climax forests similar to the Holocene mixed-oak forests in this area (West, 1984). This definition in practice can be used for the last few interglacials.

The Eemian: A Precisely Defined Interglacial

Wright (1977) pointed out a major difference between the North American and European stratigraphic records, and that observation has not changed much in

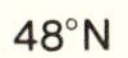

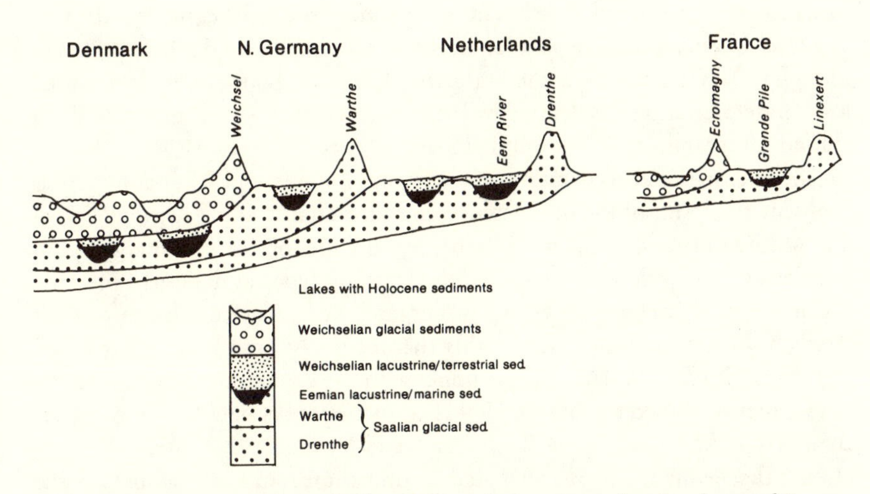

Fig. 2. Conceptual sketches of the Middle and Upper Quaternary stratigraphy of parts of Europe. Left: a cross section of the border zone of the Scandinavian ice sheet. The Eemian-type site is situated inside the outermost end moraines (Drenthe) from the next-to-last glaciation. Also the Oerel site (shown in Fig. 7) is between the Drenthe and Warthe end moraines, whereas western Holstein (Fig. 10) is between the Warthe and Weichselian moraines. The Eemian sites in Scandinavia (Figs. 1, 9, 11, and 13) are all inside the Weichselian end moraines and below Weichselian glacial sediments. Right: a similar cross section for the Vosges Mountains in northern France, included to show the morphological position of the Grande Pile site. The age assignments for the Vosges do not entirely agree with the opinions of all scientists working there.

the decade since his paper appeared. In North America very few interglacial sites have been investigated for pollen stratigraphy; in fact surprisingly few interglacial sedimentary sequences are known, and the identification of interglacials relies mainly on paleosols. In Europe interglacials are defined and named from stratotypes. Many interglacial sequences of lacustrine, terrestrial, and shallow marine sediments are known and studied. Although the definition and naming have not always been consistent and are not always generally agreed on, there is unanimous agreement on the Eemian.

The Eemian is named from the River Eem in The Netherlands, where these beds were first studied more than a century ago (Harting, 1874, cited in Zagwijn, 1961). The stratotype along the River Eem near the city of Amersfoort (Fig. 1) is documented by Zagwijn (1961). The beds were deposited in an ice-tongue basin inside the Drenthe end moraines (Fig. 2). The Eemian sequence consists of lacustrine sediments in the lower part, overlaid by marine sediments deposited during the maximum of the Eemian transgression, when the sea flooded the basin, in turn overlaid by lacustrine sediments deposited after the Late Eemian regression of the sea. The lower boundary of the Eemian was defined at a level where a subarctic park landscape was replaced by a dense forest, even though this level is missing at the stratotype. The upper boundary is placed at the level of deforestation. As defined by these boundaries, the Eem interglacial is a chronostratigraphic unit, the Eemian Stage (Mangerud et al., 1974).

The international recognition of the name Eemian and its stratotype was probably the result of the new description of the stratotype by Zagwijn. Jessen and Milthers (1928), in the most fundamental paper on the last interglacial in northern Europe, identified bogs of "the Herning type" as deposits from "the last interglacial." These bogs have two organic beds. The correlation of their stratigraphy with the one used in this chapter is easy: the lower organic bed correlates with the Eemian, and the upper with the Brørup interstadial (Fig. 3). In fact they described the Brørup Hotel bog that later became the stratotype for the Brørup (Andersen, 1961). Thus Jessen and Milthers's "last interglacial" included the Eemian, the Herning stadial, and the Brørup interstadial, in the present terminology (Fig. 3).

WAS THE EEMIAN THE LAST INTERGLACIAL IN EUROPE?

Since its discovery, the Eemian generally has been considered the last interglacial; however, it should be stressed that the stratotype defines an interglacial or more precisely a chronostratigraphic stage, whether the last or an older interglacial.

Kukla (1977) in a stimulating paper questioned whether the stratotype Eemian represents the last interglacial in Europe. He suggested that it might be the third to the last, and preferred a correlation with the oxygen isotope stage 9. Kukla's view appeared in influential textbooks such as that by Bowen (1978), gained some scientifically based popularity, and provoked a reconsideration of so-called established facts.

Most stratigraphers working in northwestern Europe disagreed with Kukla's interpretation of the Eemian (according to numerous discussions with colleagues; see also Menke and Tynni, 1984). One consequence of Kukla's interpretation was that the pollen sequence of the Eemian, so far considered diagnostic for that interglacial, must be repeated in several interglacials. Three main arguments lead to the conclusion that the Eemian represents the last interglacial.

1. The primary argument is discussed in the next section: All Eemian sites inside the Weichselian limit are overrun by ice; none outside are overridden.

2. In former lake depressions on the Saalian surface the stratigraphy is consistent: Eemian-Herning-Brørup (Fig. 3, Menke and Tynni, 1984).

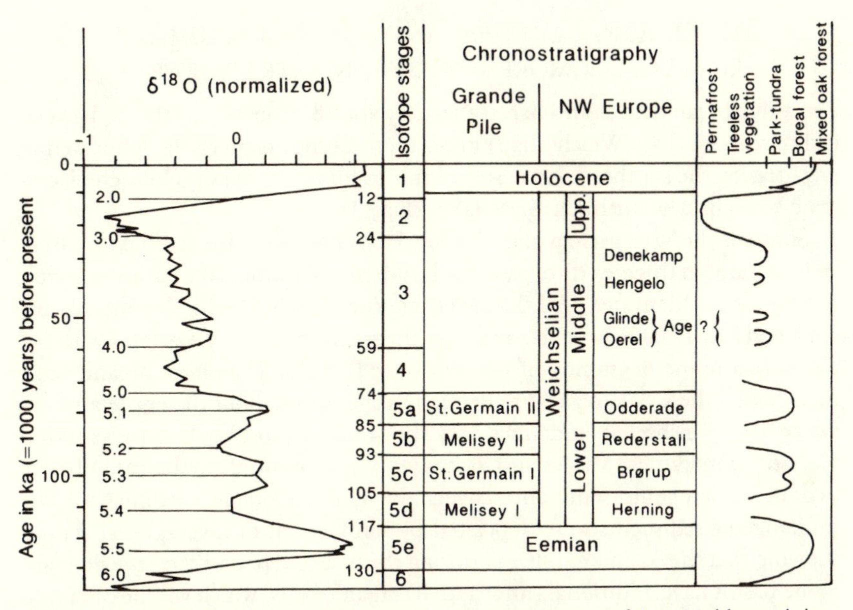

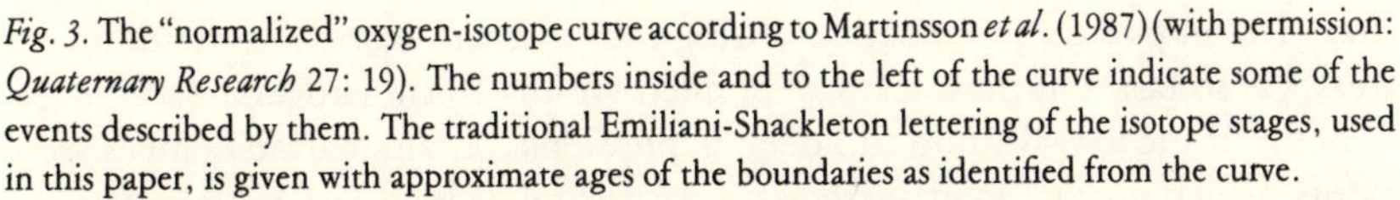

Fig. 3. The "normalized" oxygen-isotope curve according to Martinsson *et al.* (1987) (with permission: *Quaternary Research* 27: 19). The numbers inside and to the left of the curve indicate some of the events described by them. The traditional Emiliani-Shackleton lettering of the isotope stages, used in this paper, is given with approximate ages of the boundaries as identified from the curve.

The chronostratigraphy of the Grande Pile site is from Woillard (1978). The subdivision of the Weichselian into Lower, Middle, and Upper follows Mangerud *et al.* (1974), except that the middle/upper boundary is moved to around 25 ka B.P. (Chaline *et al.*, 1980). The identification and naming of the interstadials/stadials (chronozones) follow Menke and Tynni (1984) and Behre and Lade (1986). At the far right, the curve of environmental changes in Germany/The Netherlands reflects mainly the summer temperature and is based on Zagwijn (1975), Menke and Tynni (1984), and Behre and Lade (1986). Generally the warmer interstadials can be more precisely characterized than the cold, treeless periods, and I have therefore marked mainly the warm peaks. From Mangerud (in press). (With permission: Akademie der Wissenschaften und der Literatur, Mainz.)

The repetition of the Eemian pollen stratigraphy in two or three interglacials, as predicted by Kukla's interpretation, does not occur.

3. Amino-acid geochronology demonstrates that the stratotype Eemian at Amersfoort is indeed the youngest set of interglacial deposits known in Europe (Miller and Mangerud, 1986). Thus the main controversy is resolved. Miller and Mangerud also found that all sites with a typical Eemian pollen stratigraphy yielded amino-acid ratios suggesting a correlation with the stratotype Eemian, reinforcing the classical argument that the Eemian has a palynological fingerprint that can be used to distinguish it from other interglacials.

The Relation Between Interglacial Sediments, Glacial Geomorphology, and Stratigraphy

Recently glaciated areas are characterized by abundant lakes. In fact the last glacial maximum (Late Weichselian) in northern Europe is precisely delineated in any atlas by the southern and eastern limit of numerous lakes. Holocene lacustrine sequences occur in all these lakes (Fig. 2).

South of the Weichselian glacial limit, Holocene lacustrine sediments are extremely rare. In this area dozens, if not hundreds, of former lake basins not overrun by ice are identified. All, however, are completely filled in by older lacustrine and terrestrial sediments, and thus the sediment surface was level with the outlet before the beginning of the Holocene (Fig. 2). The oldest organic sediments in all these paleo-lakes are Eemian in age, throughout the entire area between the Weichselian moraines and the southernmost Saalian (next-to-last glaciation) moraines (Menke and Tynni, 1984). In most of the basins only one bed of minerogenic sediments covers the Eemian organic sediments. The minerogenic sediments were deposited by solifluction, or niveo-glacial action, showing that the basin was filled in during the first cold period after the Eemian. Some basins have additional (interstadial) organic beds, which will be discussed later. The important point is that during cold intervals erosion in the drainage areas was so fast that most lakes were filled rapidly by minerogenic sediments.

Some of the low-altitude and deep depressions contain marine Eemian sediments, which may be interlayered with lacustrine beds, for example at the Eemian stratotype at Amersfoort. Marine sediments are quite common on the Warthe tills in the coastal lowlands of northern Germany (Kosack and Lange, 1985). Marine beds (Makowska, 1982; Mojski, 1985) that Miller and Mangerud (1986) demonstrated are of Eemian age occur below the Weichselian tills in the type area for the Weichselian in Poland as well.

North of the Weichselian moraines a considerable number of Eemian sequences are also identified, both lacustrine and marine (Sjørring, 1983a). They are all covered by Weichselian till or show other signs of having been overrun by ice. At Ristinge Klint in Denmark the beds are glaciotectonically stacked, so that the Eemian beds are repeated 20 times along the cliff (Sjørring, 1983b). In some cases the Eemian sediments demonstrably overlie Saalian tills.

From the described relations (Fig. 2), we can draw the following conclusions.

1. In glaciated areas, the oldest lacustrine sediments normally date the preceding glaciation, just as the Holocene sediments date the Weichselian glaciation.

2. From point 1 it follows that the Eemian was the first interglacial after the Saalian. Even more important is the inference that if Eemian sites are securely correlated by means of their pollen stratigraphy, the Eemian was indeed the first interglacial succeeding the formation of both the outer-

most Saalian moraine (the Drenthe) and the more proximal Warthe moraines. Thus there was no interglacial between the formation of these two moraine systems.

3. All Eemian sediments inside the Weichselian limit are overrun by ice; those outside are not. Thus the Eemian must predate the Weichselian and be the interglacial between the Saalian and Weichselian glaciations. These stratigraphical relations were demonstrated clearly by Jessen and Milthers in 1928, who successfully used the occurrence of till-covered or non-till-covered deposits from the last interglacial to delineate the maximum extent of the last glaciation.

4. Glaciated terrain has limited potential for recording more than one interglacial in a lacustrine basin, even if glaciers do not overrun the basin during the intervening glaciation. The reason is that sedimentation rates during cool periods are so great that the basins would be filled in before the onset of the next interglacial.

Correlation of the Eemian Interglacial with the Deep-Sea Oxygen-Isotope Stratigraphy

This chapter has concluded that the Eemian is the last interglacial in Europe. When the Eemian is to be correlated with the oceanic record or with other continental records, however, the question is not which period in those records is considered to be the last interglacial, because that might very well be a longer period, for example, all of isotope stage 5. The relevant question is: Which stratigraphic interval in the oceanic record corresponds in time to the Eemian Stage?

In North America the last interglacial, the Sangamon, is defined from a paleosol, which is an excellent stratigraphic marker but which also has strongly time-transgressive boundaries (Johnson, 1986). According to Johnson the formation of the soil in the type area in Illinois started in isotope stage 6 and was not completed until mid-stage 3. Fulton (1984, 1986) and St. Onge (1987) use the term "Sangamon" for the entire isotope stage 5. Richmond and Fullerton (1986), on the other hand, use "Sangamon" to refer to a division of time during which the climate was similar to or warmer than the present climate, and they thus restrict the term to isotope stage 5e.

In Europe there has been a long, mainly oral, discussion on whether to correlate the Eemian with the entire stage 5 or with parts of it. At present there is general agreement that the Eemian correlates with stage 5e. The arguments for that correlation are the following.

1. This correlation was initially proposed by Shackleton (1969). His main argument is still valid. Simplified, it states that the Eemian represents one single climatic cycle, from extremely cold to warmer than present, followed by a return to cold climate in northern Europe. Thus the Eemian should correlate with a

single cycle in the ocean. It could not correlate with the entire stage 5, because that stage has three peaks. The choice of 5e, being the warmest peak in stage 5, was obvious.

2. Müller (1974) obtained a complete Eemian pollen sequence from the Bispingen site in northwestern Germany. The sediment sequence consists of lacustrine deposits, mostly of varved diatomites. Müller counted the varves and was able to estimate the duration of the Eemian as about 11 ka. He concluded that this would support Shackleton's correlation, as 5e is estimated to have lasted some 13 ka years, whereas the entire stage 5 lasted more than 55 ka (Fig. 3).

3. Zagwijn (1983) demonstrated that during the Eemian there was one single marine transgression in The Netherlands, with a regression of at least 30 m at the Eemian/Weichselian transition. This is incompatible with a correlation of the Eemian with more than one of the peaks of stage 5, because each peak was paralleled by a transgression (Chappell and Shackleton, 1986).

4. The surface-water temperature in both the eastern Norwegian Sea and along the coast of Norway is dominated by the Atlantic Current, and these temperatures are thus deemed to covary. The last time the surface water of the Norwegian Sea was as warm as during the Holocene was during stage 5e; later it was cooler (Kellogg *et al.*, 1978; Belanger, 1982). Mangerud *et al.* (1979) demonstrated that a similar warm-water pulse was restricted to the Eemian, as defined by the pollen stratigraphy of the marine sediments at Fjøsanger. Thus the Eemian is tied directly to isotope stage 5e.

5. Turon (1984) demonstrated that the Eemian correlates with 5e by pollen analysis of an isotopically dated ocean core from ca. 100 km west of Portugal (Fig. 4). In ideal circumstances, this is the most direct method of correlation. One problem with this approach is that farther from the coast, more details are lost in the pollen record, whereas closer to the coast, the isotope curve is more atypical. As Turon points out, both the pollen and the isotope curves in his study suffered slightly from that compromise.

The Last Ice Age South of the Ice Sheet

Defining and Naming the Last Glacial

The last glaciation in northern Europe is termed Weichsel or, as a chronostratigraphic unit, the Weichselian Stage, from the river that runs through Warsaw to Gdansk in Poland (Fig. 1). Weichsel is the German name of the river and is used because German geologists first described the type area along the river valley. In Polish the river is named Wisla, or as spelled in English, close to the Polish pronounciation, Vistula. Polish geologists therefore use the name Vistulian

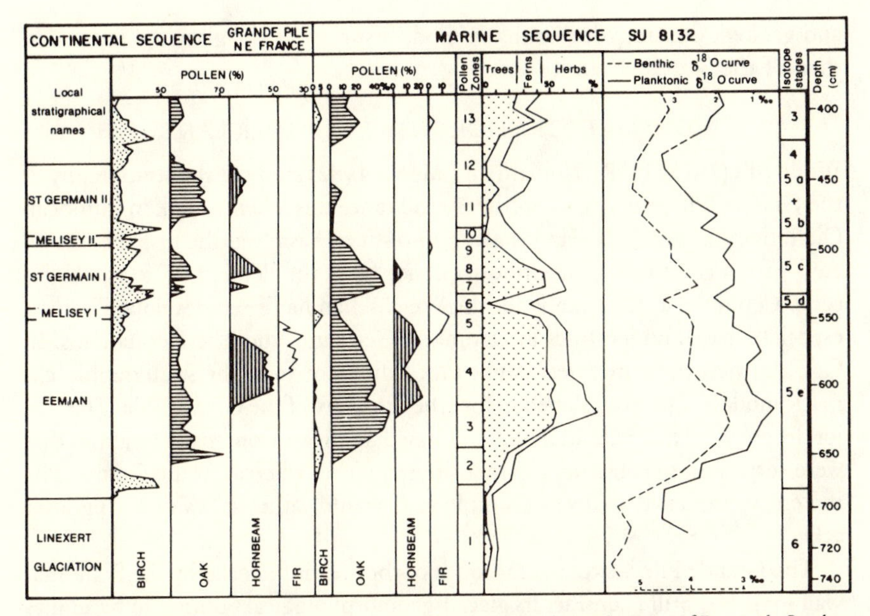

Fig. 4. Pollen and isotope curves from a core raised approximately 100 km west of Portugal. On the left is a simplified pollen diagram from Grande Pile. Turon correlated the two pollen diagrams and thus the Grande Pile record with the isotope stratigraphy. From Turon (1984). (Reprinted by permission: *Nature* 309: 673–676, copyright © 1984 Macmillan Magazines Ltd.)

(Chaline *et al.*, 1980; Mojski, 1985) and would like that to be accepted by the international community.

Most commonly the deposits of the last glaciation are easily identified in northern Europe because of the fresh end moraines and glaciogenic sediments, and nobody questions that the stratotype Weichselian represents the last glaciation. More problematic is the lower boundary of the Weichselian, because the stratotype Eemian in The Netherlands is 1,000 km from the Wisla area in Poland. This large distance is one reason that it was so difficult to prove that the Eemian immediately preceded the Weichselian.

Assuming that the Eemian was the interglacial that preceded the Weichselian, the top of the Eemian at its stratotype is accepted as the definition of the Eemian/Weichselian boundary (Chaline *et al.*, 1980). With the correlation of the Eemian with stage 5e, the Weichselian started around 117 ka B.P. (Fig. 3). The Weichselian glacial maximum occurred somewhere around 20 ka B.P. The history after the maximum is known in some detail, but it will not be described here.

According to the ages given above, the Weichselian has a history of nearly 100,000 years from its start around 117 ka B.P. up to the glacial maximum around 20 ka B.P. The main theme for the rest of this chapter is the stratigraphy and geological history during that period, first outside the glaciated areas, and then of the Scandinavian ice sheet.

Complete Stratigraphy in Former Lakes

Woillard's (1975, 1978; Woillard and Mook, 1982) study of the stratigraphy of the Grande Pile peat bog in northeastern France was a benchmark in European Quaternary stratigraphy (Fig. 5). This 20-m-deep basin was the first site discovered with a continuous lacustrine sequence from the Eemian up to the Holocene. Certainly, a stratigraphy for the Weichselian had been developed before, especially based on localities in Denmark, Germany, and The Netherlands. In fact, the view presented here is not much different from the stratigraphic/climatic model (Fig. 6) developed before the discovery of the Grande Pile. The major difference is that the old model was based on pieces from many localities that were correlated to construct a composite sequence, whereas in the Grande Pile the entire sequence occurs as a simple or (I would rather say) an exciting layer cake.

The Grande Pile is situated far south of the area influenced by the Scandinavian ice sheet. Still I consider its glacial-geomorphological position to be similar to the Eemian sites between the Weichselian and Saalian end moraines discussed above, except that the Grande Pile basin is related to moraines deposited from a local ice cap over the Vosges Mountains (Fig. 2) (Seret, 1967; Seret and Roucoux-Woillard, 1978; Dricot *et al.*, in press). The Grande Pile is situated in a valley on the southern slopes of the Vosges. Downstream of the site are the Linexert Moraines from the glaciation that formed the basin. I assume the Linexert correlates with the Saalian in northern Europe, simply because the lowest organic sediments in the Grande Pile are of indisputable Eemian age. Upstream of the Grande Pile are the Ecromagny Moraines, and even farther upstream the Servance Moraine, both of Weichselian age (Dricot *et al.*, in press).

The unique feature of Grande Pile is its topographic position; the basin has an extremely small drainage area, and thus the sedimentation rate was low enough even during cold periods that it remained a lake throughout the Weichselian (Woillard, 1978). Also the geographic position much farther south than the basins in northern Europe caused slower solifluction and other slope erosion in the surroundings during cold periods, and thus a better potential for the lake to survive.

Represented above the Eemian in Grande Pile are two warm periods that Woillard identified as true interglacials (St. Germain I and II), and she concluded that stratigraphically they should be placed between the Eemian and the Early Weichselian interstadials Brørup and Odderade of northern Europe (Andersen, 1961; Zagwijn, 1961; Averdick, 1967; see review in Menke and Tynni,

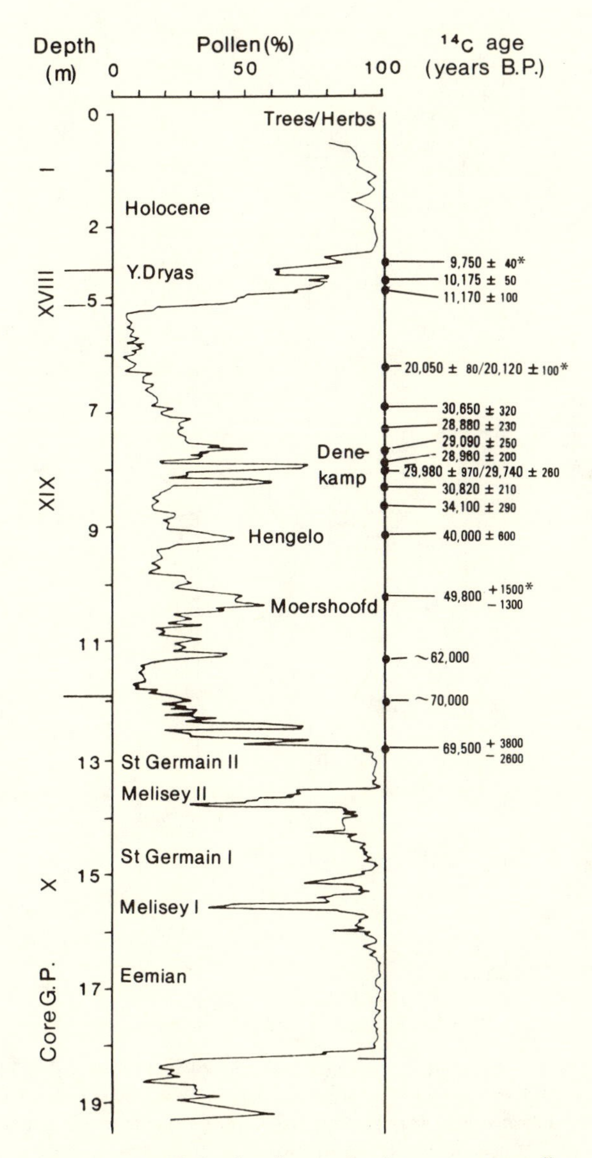

Fig. 5. A simplified pollen diagram showing percent tree pollen from Grande Pile, northern France. Names are indicated on some stratigraphic units. Dates with * are from an adjacent core. Slightly modified from Woillard and Mook (1982, with permission: *Science* 215: 159–161, copyright © 1982 by the AAAS). They concluded that St. Germain I and II probably were older than Brørup and Odderade, whereas others assume they correlate (Fig. 3).

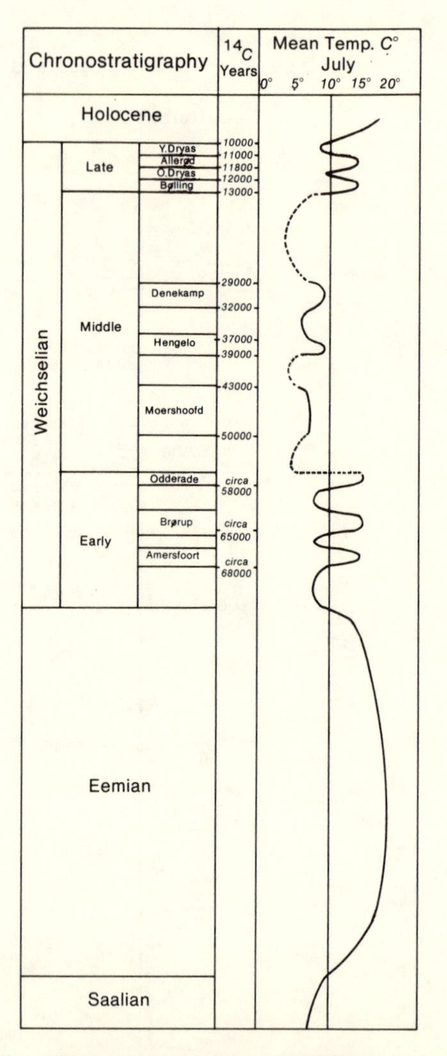

Fig. 6. A diagram slightly modified and translated from a compilation of the geology of the Netherlands by Zagwijn (1975, with permission: Geological Survey of the Netherlands). This interpretation, including the indicated ages, was representative of the view of most stratigraphers in northwest Europe at that time. The main subsequent changes in naming of units and chronological interpretations are the following (compare with Fig. 3). (1) The major change is that the Eemian is correlated with the isotope stage 5e and therefore is shorter and ends much earlier. (2) The Dutch Amersfoort and "Brørup" shown in this diagram are now thought to represent a pair of warm intervals, separated by a cooler interval, at the type site Brørup in Denmark, and are thus both included in Brørup. This has further led to a "stretching" of the Early Weichselian, and thus rejection of the radiocarbon enrichment dates that formed the base for the ages given in this diagram. (3) Most scientists now use an age of ca. 25 ka B.P. for the Middle/Late Weichselian boundary.

1984). Two main arguments led Woillard to maintain that St. Germain I and II could not be correlated with the interstadials of northern Europe.

First, St. Germain I and II palynologically show a development to a full interglacial type of mixed-oak (climax) forest that Woillard thought could not exist simultaneously with the coniferous (taiga type) forests of the Brørup and Odderade in northern Europe.

Second, Woillard and Mook (1982) obtained a radiocarbon enrichment date of close to 70 ka B.P. from the top of St. Germain II. The age of the Brørup and Odderade interstadials was assumed to be between 68 and 60 ka (Fig. 6), based on radiocarbon enrichment dates (Grootes, 1978); thus they concluded that St. Germain II was older than Brørup. What they forgot was that the Eemian always was thought to end shortly before the Brørup—for example, around 70 ka in the timescale where the Brørup started at 68 ka B.P. (Fig. 6). In other words, they moved the Eemian back in time, by accepting the correlation to isotope stage 5e, but "left the Brørup behind" by accepting the radiocarbon enrichment dates. Thus they got an open slot of time between 117 ka and 68 ka B.P., where St. Germain I and II could fit in.

Woillard's correlations have been challenged by several authors (e.g., Grüger 1979a, b; Mangerud *et al.*, 1979; Welten, 1981, 1982; Menke, 1982; Menke and Tynni, 1984; Beaulieu and Reille, 1984; Behre and Lade, 1986). They all accepted her identification of the Eemian but suggested that St. Germain I and II should indeed be correlated with the Brørup and Odderade interstadials, respectively (Fig. 3). The main argument for the latter correlation is the similar stratigraphic position. In a large number of lacustrine basins in northern Europe the first "warm-climate unit" above the Eemian is the Brørup interstadial, and in three basins that were not yet filled in during the following (Rederstall) cold period, the Odderade follows above the Brørup (Fig. 7; Menke and Tynni, 1984; Behre and Lade, 1986). It is completely improbable that any warm interstadial or interglacial should be missing between the Eemian and the Brørup in so many closed basins. Similarly, St. Germain I and II were the first warm periods after the Eemian in Grande Pile (Woillard, 1978), and this is confirmed in the basin Les Echets in France (Beaulieu and Reille, 1984).

The correlation of the Brørup and Odderade with St. Germain I and II is supported by the amplitude of environmental and climatic change. In France and northern Europe these were the two warmest periods after the Eemian, and indeed the only interstadials that were actually forested. Grande Pile is much farther south than the interstadial localities in the Netherlands, Germany, and Denmark, and therefore the Grande Pile area during warm interstadials supported a forest similar to the interglacial forests farther north in Europe. The correlation of St. Germain I and II with Brørup and Odderade implies steeper ecological and climatic north-south gradients in middle Europe than today, which is not difficult to accept. After the manuscript was accepted, an Eng-

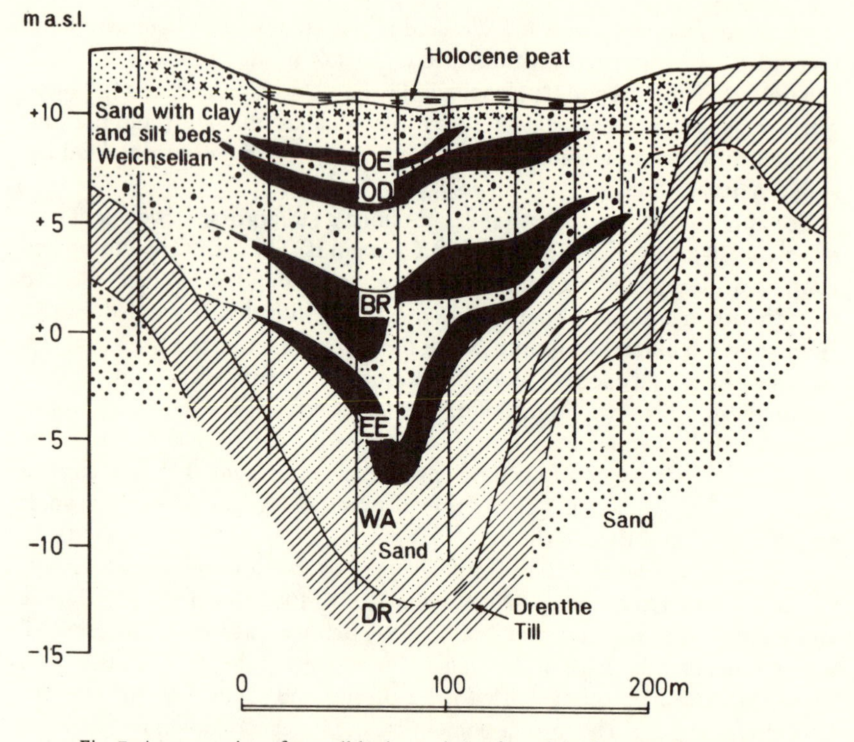

Fig. 7. A cross section of a small basin on the surface of the Drenthe till at Oerel in north Germany (Fig. 1). The solid black represents beds of interglacial and interstadial organic sediments (peat and lacustrine gyttja). On the Drenthe till the first deposits were sand, laid down during the cold Warthe (WA) phase. The first organic bed is Eemian (EE) interglacial in age. Above the Eemian are mainly minerogenic sediments deposited from solifluction and niveofluvial processes during the coldest phases of the Weichselian. However, sedimentation in that arctic environment was interrupted by the warmer Brørup (BR), Odderade (OD), and Oerel (OE) interstadials, before the basin was filled in. Slightly modified from Behre and Lade (1986). (With permission: *Eiszeitalter und Gegenwart* 36: 11–36.)

lish description (Behre, 1989) of the Oerel site (Fig. 7) appeared. Behre provides a more extensive argumentation and explanation of the correlations described above.

Correlation of the Weichselian Glaciation with the Isotope Stratigraphy

Woillard (1978) and Woillard and Mook (1982) suggested that St. Germain I and II should be correlated with the deep-sea oxygen-isotope stages 5c and 5a, respectively. To some extent this correlation was a matching of curves. In both the Grande Pile and the deep-sea these are the first warm peaks above the Eemian and 5e, and in both cases they are the highest peaks between the Eemian and the Holocene. In fact the tree-pollen curve from Grande Pile nearly mirrors the isotope curve during stages 1 through 5. Because both in some way are forced by climatic change, a correlation of the main features was all but obvious.

When the Grande Pile record is correlated with the stratigraphy in northern Germany (Fig. 3), the causative arguments for the given correlation can be further developed. At the wavelength and amplitude of change we consider here, the environmental or climatic curve for the border zone of the Scandinavian ice sheet in northern Europe must be roughly parallel to the volume change of that ice sheet. Also, at this scale, the major ice sheets on either side of the North Atlantic must covary. The discussed parallelism is demonstrated for the last glacial maximum and deglaciation. Thus at this wavelength and amplitude the climatic curve for northern Europe should approximate a northern-hemisphere glaciation curve, as does the isotope curve. Certainly, this will not be true in detail, neither concerning the timing nor the amplitude, but that is not the aim of this first-order correlation. A similar argument was used by Molfino *et al.* (1984), who stated that environments in France and the adjacent North Atlantic must have responded to the same major climatic changes. Turon (1984), in correlating the Eemian with the stage 5e, also demonstrates that the St. Germain I, and thus the Brørup, correlates with stage 5c (Fig. 4).

It follows from these correlations (Fig. 3) that the radiocarbon enrichment dates that gave ages of 60–70 ka for Odderade/Brørup provide minimum ages only. The young ages are probably a result of extremely small amounts of contamination (less than 0.5 per mil modern carbon) which can yield such ages for infinitely old samples.

The only argument that comes to mind against the proposed correlation is that it predicts marine transgressions in the North Sea area during the Brørup and Odderade, because the eustatic sea levels during stages 5a and 5c were not much lower than during 5e (Chappell and Shackleton, 1986). Such transgressions have not yet been identified.

The Western Flank of the Scandinavian Ice Sheet

The mountains in Scandinavia are like a major asymmetrical backbone from south to north, with the crest close to the western side. Thus the western side is steep with deep fjords, and it receives much precipitation from westerly winds. The eastern side is gentle and is in the rain shadow. Farther east are large plains (Finland, Estonia, etc.) and the broad depression of the Baltic Sea. The Scandinavian mountains are actually not very high, the highest summit today being 2,469 m above sea level, but they are at a high latitude, from 58° to more than 71°N.

From the topography just described one might expect that the glacial development was different on each side of the mountain chain. This is demonstrated clearly for the Younger Dryas (11 to 10 ka B.P.) when the response on the west coast was fast and large compared to more easterly areas (Mangerud, 1977). From this, it is easy to predict that future research will demonstrate more fluctuations along the west coast than the "minimum model" based on the current knowledge presented here (Fig. 8). Recent reviews of the glacial history of Scandinavia are given in Lundqvist (1986) and Mangerud (in press), where further references are given.

The following will describe in some detail three sites along the western coast of southern Norway: Fjøsanger, Bø, and Skjonghelleren. These are key sites for correlation with the stratigraphy outside the glaciated area, as discussed above, and also for generating the glacial history of the Weichselian.

FJØSANGER: UNEXPECTED INTERGLACIAL SEDIMENTS

At the Fjøsanger site, situated within the boundaries of Bergen city, Mangerud *et al.* (1979, 1981) excavated a sequence that included a complete interglacial and the earlier part of the succeeding glacial (Fig. 9). When the site was discovered (Mangerud, 1970), it was a great surprise that interglacial deposits had survived in the fjord district of western Norway, where glacial erosion has been strongest. Fjords in western Norway are up to 1,300 m deep (Sognefjorden), and outside Bergen, less than 8 km from the Fjøsanger site, the fjord is nearly 400 m deep. These fjords are carved from hard metamorphic rocks, and most have a shallow sill at the entrance. For most fjords it is clear that the overdeepening is caused by glacial erosion. Another typical feature of western Norway is that bare bedrock is exposed almost everywhere; the last glaciation removed all unconsolidated sediments, and deposited extremely limited amounts of glacial debris on land.

The Fjøsanger site is along the shore of an inlet with a sill at the mouth that is only 3 m deep. The inlet is glacially overdeepened to a depth of 90 m near Fjøsanger. No topographic or other features occur there that are different from those at thousands of other localities in western Norway. We cannot offer any explanation for why the Weichselian glaciers saved this site but cleaned the bedrock at numerous similar sites.

Basal tills that predate the interglacial occur above the bedrock (Fig. 9). The interglacial was given the local name of the Fjøsangerian and was correlated with the Eemian. The interglacial beds consist of sublittoral marine silt, sand, and gravel. Marine mollusks and foraminifera are common throughout the beds. Both fossil groups show one cycle of seawater temperature variations through the interglacial beds, from high arctic at the base to warmer than at present and then back to arctic at the close of the interglacial.

For pollen sedimentation the fjord can be compared to a large lake; however,

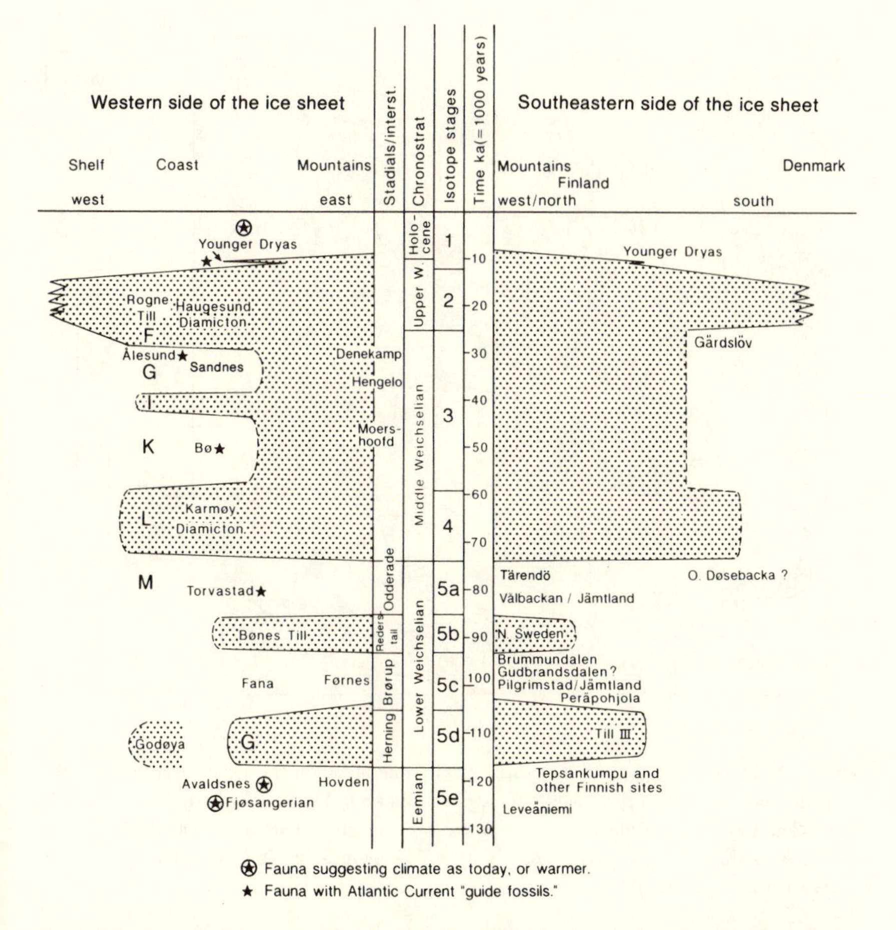

Fig. 8. Schematic glaciation curves for the last glacial (Weichselian) in Scandinavia. The curve at left is for the west side of the mountains; the curve at right, for the east side of the mountains in northern Sweden and Finland, and toward the south (Denmark) in the distal parts. The horizontal scales are somewhat arbitrary, because sites scattered over a huge area are "projected" into a theoretical, linear cross section. Ages for isotope-stage boundaries are from Martinsson *et al.* (1987; compare with Fig. 3). The indicated names are partly localities and partly stratigraphic units. The letters beneath Rogne Till are the units in Skjonghelleren (Fig. 12). Slightly modified from Mangerud (in press).

tides and stronger currents caused saw-toothed curves in the pollen diagram. One fascinating aspect of this site is that the terrestrial vegetation on the valley slopes (as inferred from the pollen record) can be directly compared to the marine fauna in the same beds. This demonstrates that the environments on land

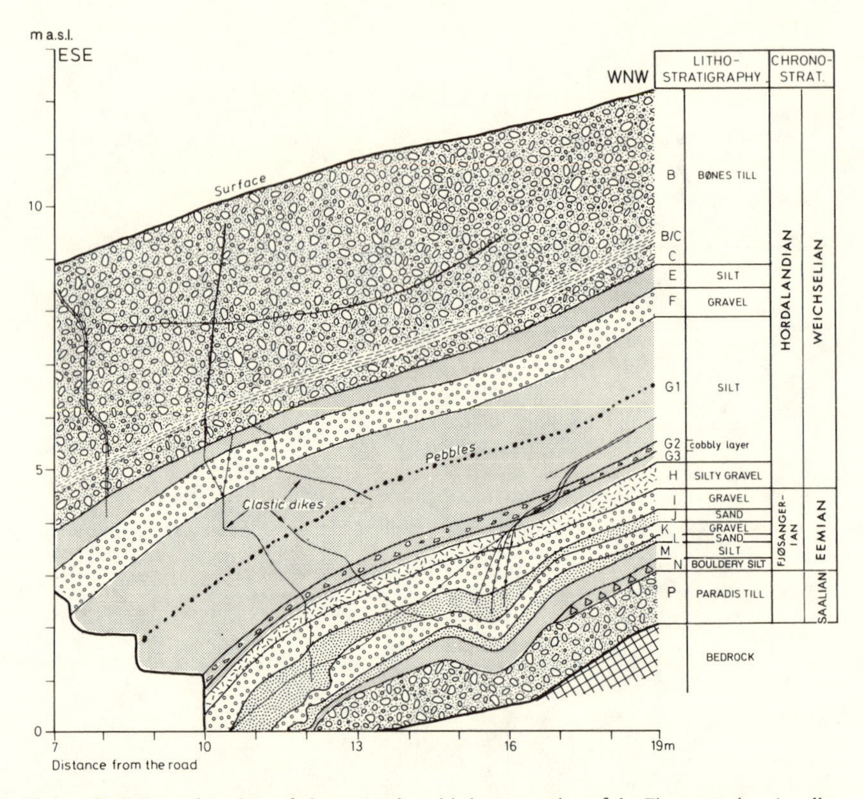

Fig. 9. The lithostratigraphy and chronostratigraphic interpretation of the Fjøsanger site. A pollen diagram from the Eemian part (beds N through I) is shown in Fig. 10. The Early Weichselian silts (beds G and E) are of glaciomarine origin, whereas the the gravel F reflects the milder Fana interstadial. Reproduced from Mangerud *et al.* (1981). (With permission: Universitets forlaget AS [Norwegian University Press].)

and in the sea were parallel, which is not surprising in western Norway where the present climate is so dominated by the warm North Atlantic Current.

Why Is the Fjøsangerian Interglacial Equivalent to the Eemian?

A main question for the glacial history after the Fjøsangerian interglacial is whether the correlation with the Eemian as stated above is correct, or if Fjøsangerian represents an older interglacial.

The principal criterion for the correlation of the Fjøsangerian with the Eemian is the pollen stratigraphy. As concluded above, in The Netherlands, Germany, and Denmark the pollen stratigraphy of the Eemian is diagnostic for that particular interglacial. The problem with Fjøsanger is that the site is far north of the classical area, and that therefore some southern, partly diagnostic trees were missing during that interglacial, as they are during the Holocene.

To make the comparison easier, three pollen diagrams are placed together to form a south-to-north profile in Fig. 10. The first is from western Holstein, just outside the Weichselian ice limit in northern Germany. The second is from lacustrine sediments at Hollerup on mid-Jutland (Fig. 1), well inside the Weichselian limit. Comparing north with south, all the important features of the Fjøsanger diagram are found in the two more southern diagrams: very distinct successional phases, early immigration of *Quercus* (oak), sparse occurrence of some other mixed-oak-forest constituents—*Ulmus* (elm); *Tilia* (linden); *Fraxinus* (ash); and a late major peak of *Picea* (spruce).

If we compare the opposite way, from south toward north, the main differences from western Holstein to Hollerup are that *Tilia* and *Abies* (fir) drop out. *Abies* is a southern conifer, thus its northern limitation is easy to understand. *Tilia,* on the other hand, thrives well in southern Norway and Sweden in the present climate, and its Eemian limit close to the border between Germany and Denmark (Andersen, 1975) is surprising.

The diagrams from Hollerup and Fjøsanger are very similar also, but two more taxa drop out: *Taxus* (yew) and *Carpinus* (ironwood). *Taxus* has its present-day northern boundary near Bergen (Fægri, 1960), but it is a poor pollen producer. More important is *Carpinus,* as the distinct *Carpinus* phase is considered a diagnostic feature of Eemian pollen diagrams from continental Europe. However, its Holocene northern limit is in southern Denmark, and even if the Eemian were slightly warmer, *Carpinus* could not be expected at Fjøsanger. The conclusion is that there is a strong palynological argument for the correlation of the Fjøsangerian with the Eemian interglacial.

Another argument for the correlation with the Eemian interglacial is the frequent occurrence of the marine gastropod *Bittium reticulatum* at Fjøsanger. This snail occurs in Europe in Holocene and Eemian deposits, but it is not reported from older sites. Thus it is considered a guide fossil for the Eemian. This argument was reinforced when Miller and Mangerud (1986) showed that all sites with *Bittium reticulatum* could be confidently correlated with the Eemian type site by amino-acid chronology, with the possible exception of Fjøsanger.

Only one type of evidence suggests that the Fjøsangerian is older than the Eemian, specifically most of the amino-acid ratios obtained by Miller and Mangerud (1986). However, some ratios support the correlation, especially the ratios from foraminifera (Miller *et al.*, 1983; Sejrup, 1987). Sejrup (1985, 1987) has also demonstrated that some of the mollusks at Fjøsanger have not acted as closed systems for amino acids.

At Fjøsanger we also have tried thermoluminescence dating. The first results clearly suggested an Eemian age (Hütt *et al.*, 1983). The major problem with

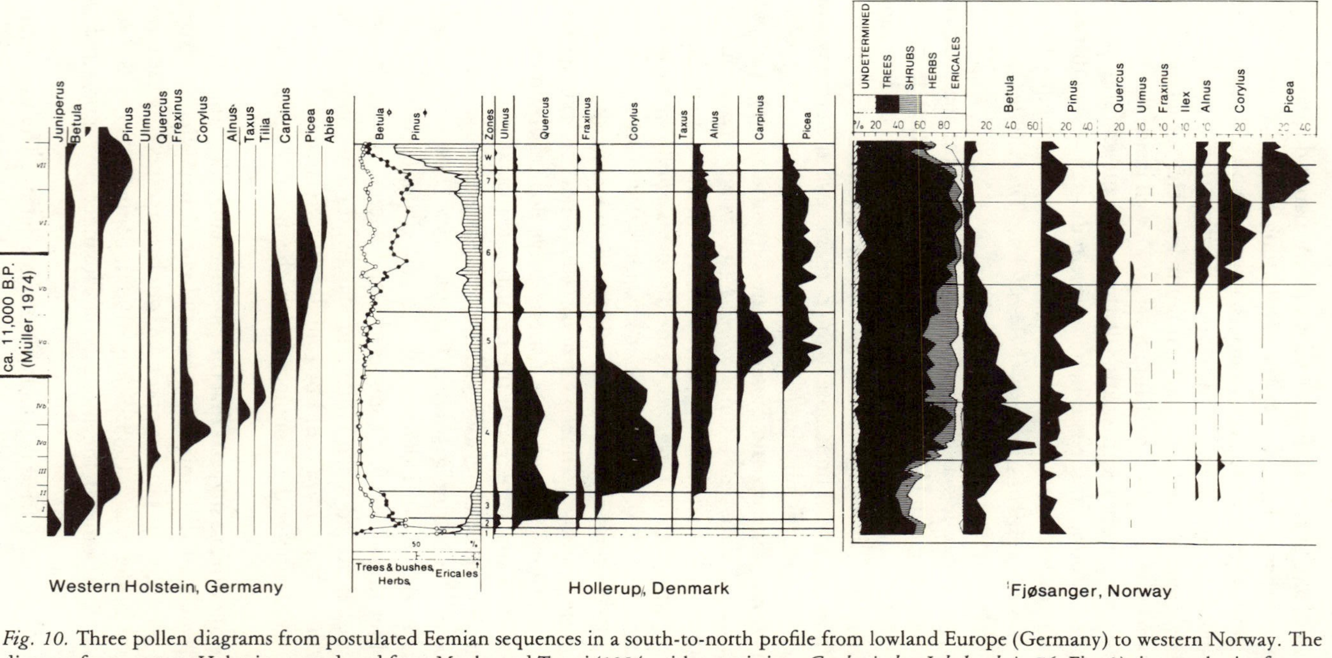

Fig. 10. Three pollen diagrams from postulated Eemian sequences in a south-to-north profile from lowland Europe (Germany) to western Norway. The diagram from western Holstein, reproduced from Menke and Tynni (1984, with permission: *Geologisches Jahrbuch* A, 76, Fig. 3), is a synthesis of many diagrams from lacustrine sediments and thus shows an idealized pollen sequence from that area. Note also that it is plotted on a linear timescale, based on varve counts by Müller (1974). The two other diagrams are from single localities, and the vertical scales are sediment thickness. The paleolake at Hollerup had slow sedimentation in the early part, thus the *Corylus* increase occurs low in the diagram (from Andersen, 1965). At Fjøsanger the opposite was the case, thus the *Corylus* increase is high up, and the *Picea* zone is thin (from Mangerud *et al.*, 1981). (With permission: Universitets forlaget AS [Norwegian University Press].)

those samples was that they had been collected long before being dated, so we used the central parts of blocks of sediments, assuming light can penetrate only into the very outer crust. Recently we dated samples collected for the purpose of thermoluminescence dating and also used improved dating techniques. The preliminary results confirm an Eemian maximum age (Mejdahl and Mangerud, unpublished).

I find the arguments favoring the correlation of the Fjøsangerian with the Eemian are convincing enough to conclude that the correlation is correct, and for the rest of this chapter that is assumed.

The Start of the Last Glaciation: Isotope Stages 5d and 5c

The first sign of a coming glaciation at Fjøsanger is the initial accumulation of ice in North America or Antarctica. During the later phases of the interglacial, sea level dropped some 10 to 20 m, probably because of eustatic sea-level lowering caused by ice accumulation (Mangerud *et al.*, 1979; in press). At that time there were still thermophilous forests at Fjøsanger, so the possibility of any ice growth in Scandinavia can be eliminated.

Just above the interglacial beds is a glaciomarine silt 2.5 m thick (G in Fig. 9) that demonstrates that glaciers terminated in the fjord some few km from the site. Several observations suggest that no major hiatus exists between the interglacial and the deposition of the glaciomarine silt G: The beds conformably overlie each other in several different excavations, with different topographic positions. Pollen, foraminifera, and amino acids also suggest continuous deposition. The conclusion from these observations is that the glaciomarine silt (bed G) was deposited soon after the Eemian, which means that its age is isotope stage 5d. Thus we can demonstrate a glacier in western Norway during isotope stage 5d that in the Bergen area approached the size of the Younger Dryas ice.

The next question is the duration of this glaciation, and that is a more difficult problem. Gravel F, above glaciomarine silt G, contains rocks of local provenance and abundant mollusks. Both observations show that the glacier had withdrawn from the fjord. Mangerud *et al.* (1981) thought that this warming was a local event, named the Fana interstadial, and was simply a result of the ice front retreating some few km, enough that meltwater no longer entered the fjord. The argument was that the fauna in gravel F does not contain any warm-water element—for example, no species that are restricted today to the Atlantic water. Thus Fana was colder than expected if it were to be correlated with the Brørup. With these assumptions the Fana, and also the Bønes Till capping the Fjøsanger section (Fig. 9), may be isotope stage 5d in age, though I have recently favored the following alternative (Mangerud, in press).

Amino-acid analysis indicates that the Fana interstadial has an age slightly younger than 100 ka B.P., and thus it should be correlated with isotope stage 5c (Miller *et al.*, 1983; Sejrup, 1987), and indeed with the Brørup on the continent (Fig. 3). If we accept this correlation, the unexpected cool fauna can be explained in two ways: (1) the warmest part of the interstadial is missing; (2) the difference in summer temperature between the present and the Brørup was larger in western Norway than farther east in Europe. The latter explanation is supported by temperature gradients deduced from pollen sequences in Europe, and I assume it is at least partly true if the correlation between Fana and Brørup should prove correct.

The difference between the two alternatives may be considered minor concerning the absolute age of the Fana interstadial. However, for the glacial history the implications are significant. If Fana correlates with isotope stage 5c, then the entire stage 5d should be found in silt G. According to Chappell and Shackleton (1986) the sea level dropped 70 m from peak 5e to the trough of 5d. The only way to assume continuous sublittoral deposition at Fjøsanger, during a period with such substantial fall in eustatic sea level, is to postulate a glacio-isostatic depression that compensated for the drop in sea level. This would require a considerable size for the Scandinavian ice sheet, which is indeed compatible with the conclusion that it approached the size of the Younger Dryas in the Fjøsanger area.

The Bønes Till: Isotope Stage 5b?

Above the gravel F is the glaciomarine silt E, showing that glaciers again approached Fjøsanger. Silt E is overlaid by the Bønes Till, whose age was discussed briefly above. Three observations have led to the conclusion that no unconformity exists between glaciomarine silt E and the Bønes Till (Fig. 9). (1) In three of four excavations the till conformably overlies the silt E, which is 0.5 m thick. It would be most surprising if glacial erosion had stopped at several different places at just this stratigraphic level. (2) Upglacier, in the deeper part of the fjord, the glacier eroded marine sediments that it overrode, and therefore the till is full of transported fossils. These fossils are correlated with the Eemian and Early Weichselian beds at Fjøsanger; there are no hints of younger fossils. (3) A large number of amino-acid analyses of shell fragments in the till gave ratios corresponding only to the beds at Fjøsanger, suggesting that the glacier did not erode any younger beds.

Point 2 is also a strong argument that the Bønes Till was deposited by the first glacier that overrode the site. The important message for the age of that glaciation is that it occurred soon after Fana (gravel F). Thus if Fana is stage 5c/Brørup, the Bønes probably is stage 5b. Because no sediments are found above the till, a minimum age cannot be proved.

Bø on Karmøy: One Interglacial and More Interstadials

Karmøy is a low island facing the North Sea. In excavations in the former clay pit at Bø, Andersen *et al.* (1983) reached a sand that they named the Avaldsnes

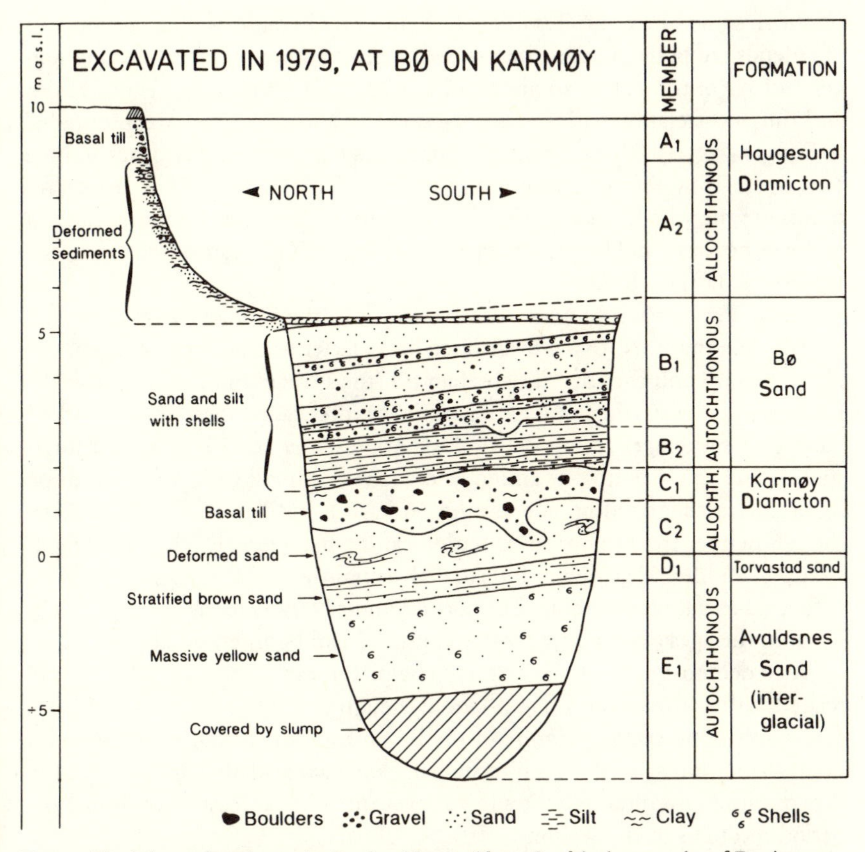

Fig. 11. The lithostratigraphy at the Bø site. The Avaldsnes Sand is shown to be of Eemian age. Amino-acid ratios have demonstrated a major hiatus between the Avaldsnes and the Torvastad. The Karmøy and Haugesund diamictons were deposited by ice advances during isotope stages 4 and 2, respectively (Fig. 8). The site is exceptional because sediments from so many events are preserved in a heavily glaciated area. Slightly modified from Andersen *et al.* (1983, with permission: Geological Survey of Norway).

Sand (Fig. 11). The pollen, mollusk, and foraminifera stratigraphy are so similar to the Fjøsangerian record that a correlation hardly can be questioned (Andersen *et al.*, 1983, oral communication 1988; Sejrup 1987). The only difference is that the amino-acid geochronology provides straightforward support for the correlation of the Alvaldsnes interglacial with the Eemian (Miller and Mangerud, 1986; Sejrup 1987). Thus an Eemian age of this interglacial is quite certain.

Directly on the interglacial beds lies the Torvastad (interstadial) Sand, where the marine fossils suggest conditions comparable to the northernmost tip of Norway today (Sejrup, 1987). The pollen assemblage indicates an open vegetation with some birch, but it is difficult to interpret because of the large amount of redeposited pollen in these marine sediments. Amino-acid racemization suggests an age of the Torvastad interstadial of 78 ± 7 ka B.P. (Miller *et al.*, 1983), and thus a correlation with the isotope stage 5a and the Odderade interstadial in Europe (Fig. 3). This suggests a major hiatus between the Avaldsnes and the Torvastad. How this unconformity was formed is unknown. The important points are that the fauna suggests that a branch of the North Atlantic Current reached the coast, and that this happened during the Odderade interstadial/isotope stage 5a (Fig. 8).

Above the Torvastad interstadial is a basal till (the Karmøy Diamicton, Fig. 11) showing that the site was overridden by a glacier reaching the open sea. Between this till and the Late Weichselian till (the Haugesund Diamicton) is the Bø Sand, demonstrating another ice-free (Bø) interstadial. Several radiocarbon dates gave finite ages around 50 ka B.P. (Andersen *et al.*, 1983) for this interstadial, an age supported by amino-acid analyses of mollusks. Amino acids on foraminifera suggested an age around 60 ka B.P. (Miller *et al.*, 1983). Even though neither the age nor the duration can be fixed exactly, the site demonstrates a Middle Weichselian interstadial, certainly predating ca. 40 ka B.P., with a sea-surface temperature like northernmost Norway today (Sejrup, 1987).

From the ages given above for the Torvastad and Bø interstadials, the glacial advance demonstrated by the Karmøy Diamicton can most reasonably be correlated with isotope stage 4 (Fig. 8), even though that is a correlation with weak constraints. One exciting feature of the Bø on Karmøy section is that so many events—one interglacial, two interstadials, and two stadials—are preserved in superposition, even though the site was overrun by ice at least twice during the period recorded in the section.

SKJONGHELLEREN: A CAVE WITH A UNIQUE RECORD OF GLACIATIONS

Skjonghelleren is a wave-cut cave 100 m long (Fig. 12a), formed in a vertical bedrock cliff. From the cave opening one has a view toward the open ocean. The cave is positioned well above the postglacial marine limit; thus it must predate the last glaciation of the site. The stratigraphy in the cave shows alternating beds reflecting two sedimentary environments (Fig. 12b). (1) When glaciers passed the cave mouth, a lake was dammed inside the cave, and thick sterile beds of finely laminated clay were deposited in the lake. (2) During ice-free conditions, blocks were weathered from the roof, speleothems were precipitated, and bones were brought into the cave by humans or other animals.

These two groups of sediments were discovered and correctly interpreted by the famous Norwegian geologist H. H. Reusch (1913). The sequence became important for the history of the last glaciation through deep excavations and borings by Larsen *et al.* (1987), from which the information in this paper is taken.

The sediment sequence in the cave proved to be about 20 m thick (Fig. 12a), admittedly much more than was thought before starting the excavation. There are three beds with laminated clay, the thickest around 4 m, and four blocky beds, including the Holocene on top. Thus the sequence records three glaciations with ice-free periods before, between, and after the glaciations. Probably the entire sequence is Weichselian in age, even though that is not proved. In Fig. 8 the glaciations are plotted according to the best age estimates.

The discussion will concentrate on the last ice-free period, because that is the only one that is dated confidently. The corresponding blocky bed was described as "diamicton G" because of a silt matrix mixed in by solifluction. Nearly 7,000 bones were unearthed from this bed. Two radiocarbon dates on bones and three U/Th dates on speleothems from the bed all yielded ages close to 30 ka B.P.

When this last ice-free period started and its duration are uncertain. In time, it corresponds to the Ålesund interstadial, earlier defined in this area on the basis of dated fossils in till. The radiocarbon dates reach back to around 38 ka B.P., but they may stem from different ice-free periods. Paleomagnetic excursions in the underlying clay bed (Bed I) (Fig. 12b) are possibly correlatable with the Laschamp/Olby event, dated at between 36 and 42 ka B.P. Probably the Ålesund interstadial lasted a few thousand years.

The age of the last ice advance, and thus the end of the Ålesund interstadial, can be better dated. Apparently this cave offers a rare opportunity to precisely date an ice advance. The advance has to postdate the radiometric dates of 28–30 ka B.P. In the overlying glaciolacustrine clay (bed F), the virtual geomagnetic pole defines two loops, close to the equator and between 60° and 90°E. This is best correlated with the Lake Mungo event, dated at 28–30 ka B.P. With the available information, we can conclude that the glacier front passed Skjonghelleren close to 28 ka B.P. on its way to the Late Weichselian maximum.

Skjonghelleren is only about 65 km from the continental edge, which certainly is an outer limit for grounded ice. The area around Skjonghelleren was not deglaciated again until around 12.5 ka B.P., suggesting that the ice front in this area remained near its maximum position for 15,000 years, from 28 ka to 12.5 ka B.P. (Fig. 8).

The faunal record of the Ålesund interstadial provides valuable information about the environment around 30 ka B.P. Most of the bones are from birds that nest in colonies in northern Norway and Svalbard, suggesting that the cliff into which Skjonghelleren is developed was a nesting cliff during the Ålesund. Other indicator fauna include the fish otter (*Lutra lutra*) and fish species *Pollachius virens* and *Brosme brosme* (pollock and cusk). These taxa presently are not found north or east of northernmost Norway. This means that they do not survive in Arctic waters but are restricted to the North Atlantic Current. Thus their occur-

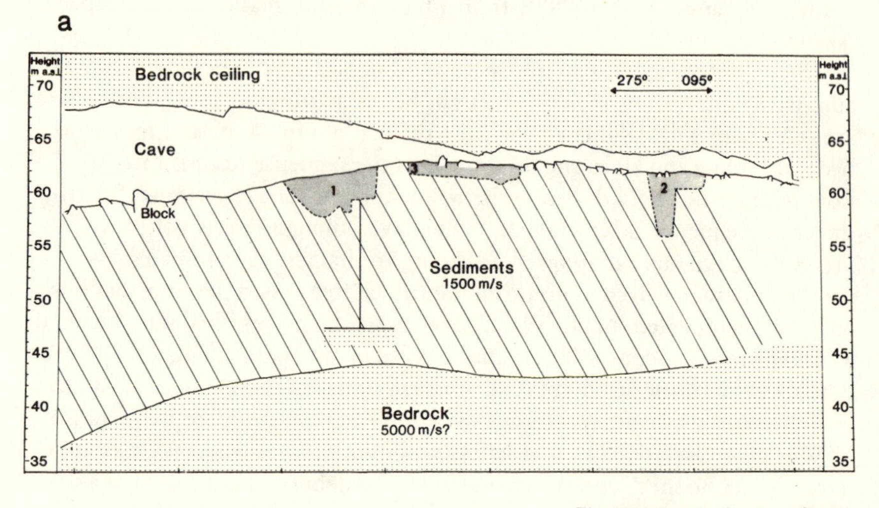

Fig. 12a. Longitudinal profile of the Skjonghelleren cave. Note that the cave in the inner part (to the right) today is only 2–3 m high, whereas the sediments are more than 15 m thick. The excavations are marked 1, 2, and 3. The deepest coring from excavation 1 is marked; it showed slightly less depth than suggested from the seismic profiling. Reproduced from Larsen *et al.*, 1987. (With permission: Universitets forlaget AS [Norwegian University Press].)

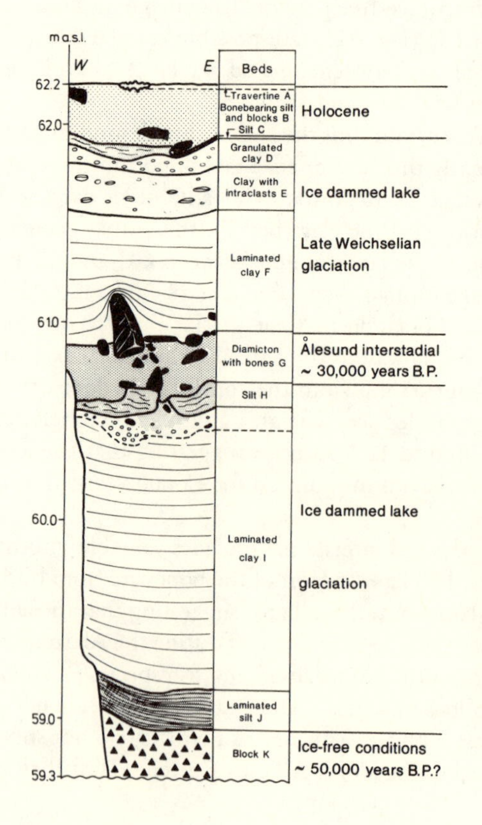

Fig. 12b. Stratigraphy in the upper part of excavation 2. The lowest shown bed, block K, is 3 m thick. The coring showed that beneath bed K is another bed of laminated clay (L on Fig. 8) and a bed of blocks (bed M) on the bedrock. Slightly modified from Larsen *et al.*, 1987. (With permission: Universitets forlaget AS [Norwegian University Press].)

rence suggests that a branch of the North Atlantic Current swung into the Norwegian Sea during the Ålesund interstadial.

The Core Area of the Ice Sheet

The central area of the former ice sheet obviously is the place to investigate how far back the ice retreated during the interstadials. That problem is discussed only briefly here. For recent reviews refer to Lundqvist (1986) and Mangerud (in press).

Many interstadial sites are known in the central area of Norway, Sweden, and Finland. The major problem is to date and correlate the sequences. In most cases it is impossible even to prove that they are of Weichselian age, though it is hard to conceive that many of them should be even older.

NORTHERN FINLAND: MANY SITES, ONLY ONE INTERSTADIAL

In northern Finland I find it proven beyond doubt that at least one ice-free interstadial occurred. This was first demonstrated by Korpela (1969), who named the interstadial Peräpohjola from the district where it was discovered. Recent reviews are given by Hirvas *et al.* (1981) and Hirvas and Nenonen (1987), from where most of the following is cited.

At several places in northern Finland peat beds beneath till occur with a pollen flora that demonstrates that the climate was warmer than or as warm as present (Hirvas and Kujansuu, 1981). An interglacial age on that basis is unquestionable. The correlation to Germany and The Netherlands is difficult because the location is remote and few species of trees grow that far north. Thus a comparison of the pollen stratigraphy, as discussed earlier for the Fjøsanger site, is at the moment impossible, though that may improve in the future, with a denser network of interglacial sites. Most of the sites in Finland are assumed to be of Eemian age, which is certainly reasonable.

Above the youngest interglacial beds is a till bed (Till III, Fig. 13) that is laterally correlated over large areas, mainly by means of till fabric. At many sites organic sediments are found on tills identified as till bed III, though never in direct superposition on a site with interglacial peat. The pollen flora in these sediments is always dominated by *Betula* (mainly tree-birch type) and herbs, suggesting a birch forest and thus cooler conditions than during the Holocene. All these sites are referred to the Peräpohjola interstadial (Fig. 8).

The conclusion of Hirvas *et al.* (1981), Hirvas and Nenonen (1987), and many earlier authors cited by them is that during the Weichselian there were two glaciations in northern Finland separated by one single ice-free period, the Peräpohjola interstadial, correlated with the Brørup. The strength of this model is that it apparently explains all observations. Most important, it has been

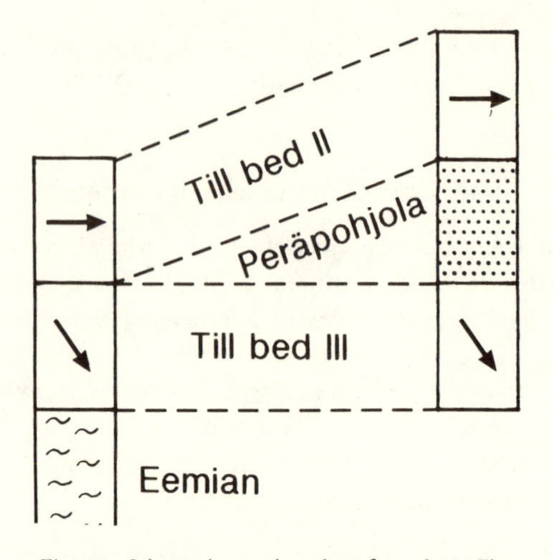

Fig. 13. Schematic stratigraphy of northern Finland. Arrows show ice-flow directions as indicated by till fabric, which is the main tool for stratigraphical correlations in that area. Interglacial and interstadial sediments are not yet discovered in superposition in the same section. On the other hand, the relationships shown are completely consistent; Peräpohjola interstadial sediments never occur beneath a till with southeasterly fabric, or above one with fabric due east.

demonstrated stratigraphically that at least one ice-free interstadial occurred after an interglacial that hardly could be anything but the Eemian.

Concerning the age of the Peräpohjola, refer to the discussion above, on the Weichselian outside the glacial limit. Compared to northern Germany, it is clear that the only periods during the Weichselian that had a climate warm enough to allow birch forests to grow in northern Finland were the Brørup and Odderade interstadials (Fig. 3). That correlation means a minimum age of 74 ka B.P., which in turn suggests that the many samples from the Peräpohjola yielding finite radiocarbon ages have been contaminated.

The Nucleus of the Southern Part of the Ice Sheet

Many sites with sediments beneath till, most with only minerogenic sediments, are known from areas close to the drainage divide in the southern parts of the Scandinavian mountains (e.g., Lundqvist, 1967; Bergersen and Garnes, 1981). When these sites were ice-free, nearly all glacier ice in Scandinavia must have been gone.

Two key sites are Brumunddal (Helle *et al.*, 1981) and Pilgrimstad (Lundqvist, 1967) because they include organic sediments that offer the possibility of pollen stratigraphy. At Brumunddal (Helle *et al.*, 1981) the lithostratigraphy (till/peat/till) and the pollen stratigraphy show a full environmental or climatic cycle: glaciation/tundra/birch forest/tundra/glaciation, in which the ice-free period did not reach the temperature of an interglacial. The frequent occurrence of *Larix* (larch) and *Picea* within the *Betula* forest suggests a correlation with the Brørup interstadial, even though both trees also occurred during the Odderade in northern Europe. The underlying till suggests a glaciation between the preceding interglacial and the Brumunddal interstadial. The pollen diagram from Pilgrimstad (Lundqvist, 1967) is very similar to that from Brumunddal, except that *Larix* is missing, and it probably correlates with the same interstadial.

Another interesting site is Vålbacken, near Pilgrimstad, where Lundqvist (1967) described clay and silt beneath till. Lundqvist included these sediments in the Jämtland interstadial, which he correlated with the Brørup interstadial. Later Mörner (1981) found that the Vålbacken sediments have a paleomagnetic signal similar to the signal in St. Germain II in the Grande Pile and different from the signal in St. Germain I. The correlations given in Fig. 8 would mean that Vålbacken is of Odderade age. Thus if the commonly accepted correlation of some of the other sites with Brørup is correct, at least two ice-free periods have occurred in the central areas.

Northern Sweden—New Results: Two Interstadials in Superposition

Some of the most challenging interpretations of glacial landscapes and the history of the last glaciation in Scandinavia have recently come from studies by Lagerbäck (1988a, b; Lagerbäck and Robertsson, 1988). In one area in northeastern Sweden (Fig. 14) the landscape is dominated by an older system of drumlins, formed by an ice movement from the northwest. These drumlins are generally 1–3 km long, 200–500 m across, and 10–20 m high. Nearly parallel with the drumlins run large, sharp-crested eskers, several tens of km long.

In parts of the area younger drumlinoid forms occur, mainly as superimposed flutings on top of the older drumlins. These forms indicate ice movements from the south or southwest (Fig. 14). In addition to the two main systems, more westerly drumlins occur locally; their age relation to the main systems is not clear.

Along the northwestern–southeastern (older) eskers, kettle holes are frequent. Lagerbäck proposed that these kettle holes offered good conditions for deposition and preservation of organic matter in possible ice-free periods, and thus he cored and excavated some of them. The stratigraphy of a site named Onttoharjut is shown in Fig. 15. Nearly identical stratigraphy, with a till bed between two organic beds, was found at four localities (Lagerbäck and Robertsson 1988), and this sequence is considered representative for the history of the area. This stratigraphy is supported by some additional 25 sites with one organic bed that in most cases can be correlated with one of the beds at Onttoharjut by means of the glacial sediments beneath or above the organic bed.

The following outlines the history of the area according to the interpretations of Lagerbäck and Robertsson (1988) and Lagerbäck (1988 a,b), using the stratigraphy on Fig. 15 as an example.

The lowest unit is the esker gravel, and thus the northwesterly esker proper. The thick till (and drumlins, Fig. 14) belonging to this phase can be correlated with till bed III in Finland (Fig. 13) and also with a till between the Eemian and some overlying (interstadial) water-laid sediments at Leveäniemi described by Lundqvist (1971). The conclusion is that these sediments and landforms were formed during the first glaciation after the Eemian, isotope stage 5d (Fig. 8).

On the gravel is a peat (Fig. 15) for which most radiocarbon dates yielded infinite ages. The pollen composition demonstrates that during the climatic optimum an open park tundra existed with birch (*Betula pubescens*) as the only tree. Maximum values of *Betula* pollen are around 50%. Thus the climate was colder than at present, as today the area is forested by conifers (*Picea* and *Pinus*). The pollen stratigraphy correlates well with the Peräpohjola interstadial in Finland, which is assumed to correlate with the Brørup interstadial farther south (Fig. 8).

Above the peat is a till deposited by an ice sheet that overran the area. The ice incorporated some of the organic bed into the till but did not alter the morphology of the eskers or drumlins to any degree. Till fabrics generally indicate ice flow from the west. Thus the till cannot be correlated with either of the two major systems of drumlins (Fig.14), but it may correlate with the more locally occurring westerly drumlins mentioned above. The till is interpreted to correlate with isotope stage 5b (Fig. 8).

The laminated organic sand represents a second ice-free period. This is the first time that two superimposed interstadials have been demonstrated in central areas of the former Scandinavian ice sheet. The period is named the Tärendö interstadial from a village in the area. From this interstadial also, many radiocarbon dates yielded infinite ages, and samples with finite ages are assumed to be contaminated. The pollen assemblage indicates colder and drier conditions than during the first interstadial, but still with a considerable amount of pollen of birch trees during the optimum. Ventifacts (wind-abraded stones), ice-wedge polygons, and frost-shattered bedrock found on the present surface in restricted areas are correlated with this interstadial, implying that these areas escaped erosion during the subsequent glaciation. The Tärendö interstadial is correlated with the Odderade (Fig. 8), which was the last time before 13 ka B.P. that forests grew in northern Germany and thus the last time trees could have grown in northern Scandinavia.

Capping the sand is a till-like sediment (Fig. 15) correlated with till in the southern youngest drumlin system (Fig. 14). According to the ages assigned in

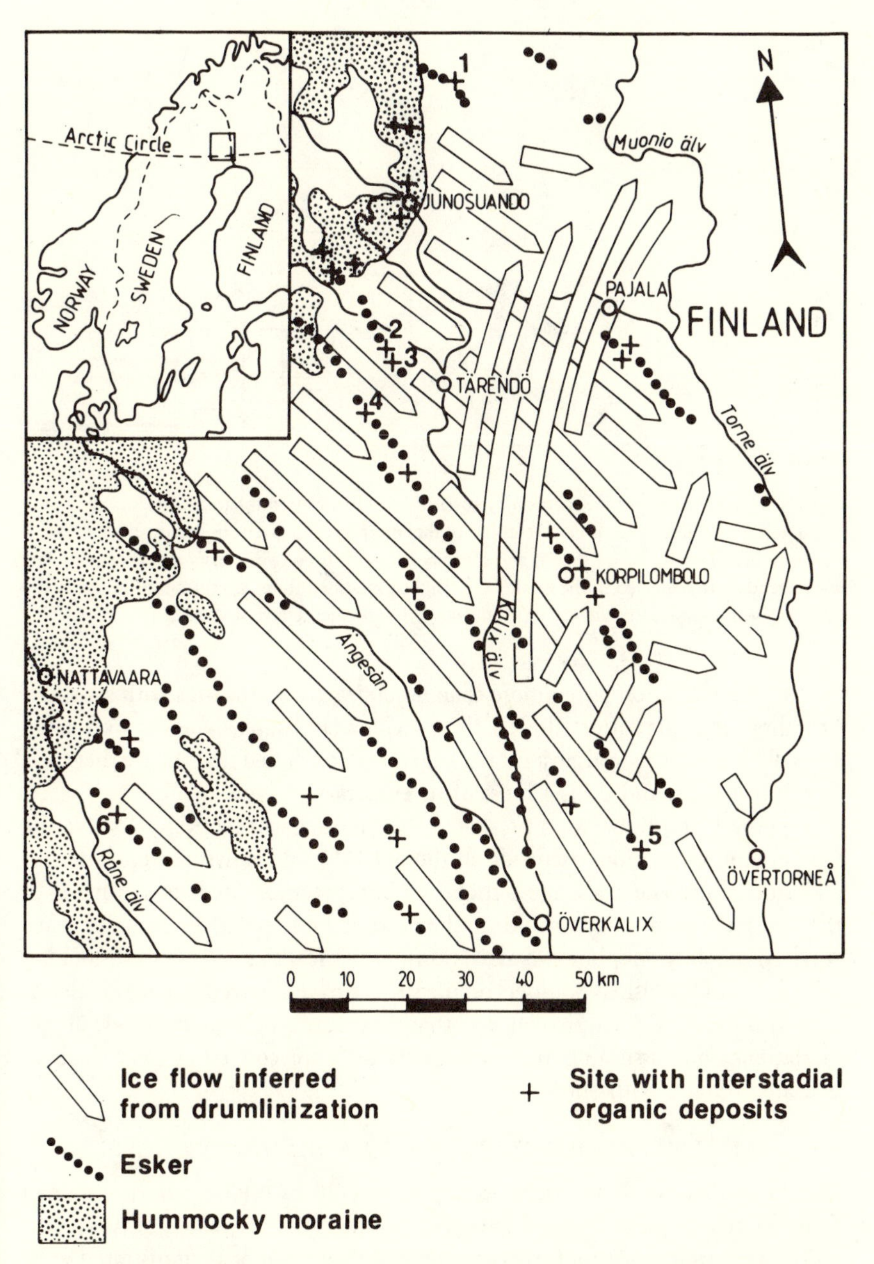

Fig. 14. The glacial geology of an area in northeastern Sweden (see inset). Only the ice-flow directions (drumlins) toward N and NE are assumed to be of Late Weichselian age. The major drumlin and esker system toward SE is dated to Early Weichselian. Reproduced from Lagerbäck and Robertsson (1988). (With permission: Universitets forlaget AS [Norwegian University Press].)

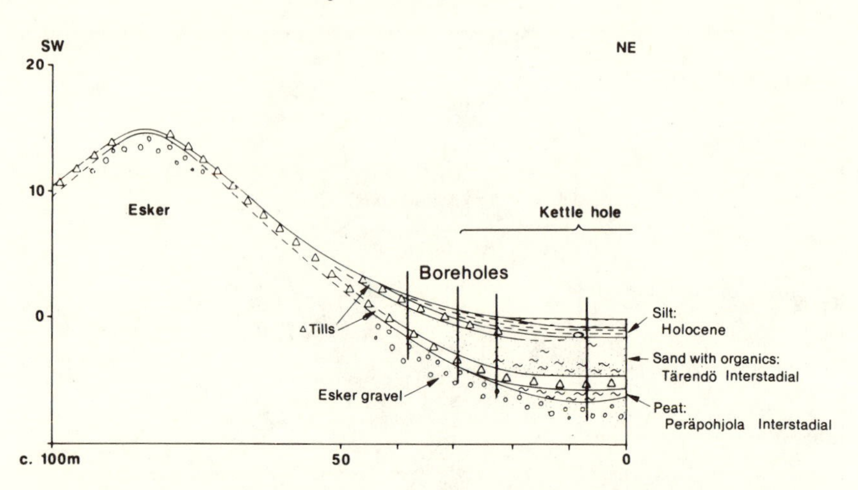

Fig. 15. A cross section of an esker running toward the southeast (Onttoharjut, site 2 in Fig. 14), and a kettle hole along the esker. According to the interpretation given, this sharp-ridged esker has twice been overrun by ice sheets, and organic sediments from two interstadials (Peräpohjola and Tärendö) are preserved in the kettle hole. Slightly modified from Lagerbäck and Robertsson (1988). (With permission: Universitets forlaget AS [Norwegian University Press].)

Fig. 8, that glaciation lasted more than 60,000 years. In the area with southern drumlins, that glaciation altered the previous landscape, whereas over large areas older deposits are almost intact. Lagerbäck concluded that the ice sheet was frozen to the ground during most of its existence.

Lagerbäck and Robertsson presented their results on a NORDQUA (Nordic Quaternary Association) excursion in the fall 1987 and convinced me, but not all participants. The challenge is that their interpretation predicts that much of the landscape seen today, including sharp-crested eskers and kettle holes, have survived not only one, but two glaciations. Other features such as fresh-looking talus and wind-polished boulders on the surface have survived one major glaciation. It is indeed difficult to conceive that these forms have survived one or two glaciations, but assuming their interpretations are correct, we can look at glaciated terrains with "new eyes."

ICE-FREE PERIODS WITH FINITE RADIOCARBON AGES?

Many finite radiocarbon dates with ages above 20 ka B.P.have been obtained from the central parts of Scandinavia (see Lundqvist, 1981, 1986; Hirvas *et al.*, 1981; Mangerud, 1981 for further references). Many can be demonstrated to be minimum ages only, for example, because they date forested periods. In my opinion no finite date from central Scandinavia, that is, the area proximal to the Younger Dryas end moraines, can at present be convincingly argued to be correct. This may change in the future, but in Fig. 8 it is assumed that the central areas were covered by ice from 50 ka to 12 ka B.P.

Bergersen and Thoresen (oral communication, 1988) recently obtained a thermoluminescence date of 37 ka B.P. for windblown sediments from Gudbrandsdalen (Fig. 1). Lauritzen (oral communication, 1988) obtained U-series dates of around 30 ka B.P. on speleothems from Nordland (Fig. 1). Both cases have to be retested, and I have been conservative by not including them in the diagram (Fig. 8). Still the methods used open exciting ways of dating ice-free periods too cold, or too short, for organic production. If they are correct, each would demonstrate that central areas were indeed ice-free.

Summary

The Eemian, as defined from the stratotype in The Netherlands, was the last interglacial in Europe. The Eemian interglacial correlates with isotope stage 5e (Fig. 3). During the Early Weichselian two interstadials occurred in northwestern Europe, the Brørup and Odderade, both only a couple of degrees cooler than the present climate. They are correlated with isotope stages 5c and 5a, respectively (Fig. 3). During both interstadials, forests grew in northern Germany. From 75 ka B.P. to 13 ka B.P., there were no forests in northern Germany, and thus none in Scandinavia.

The growth of the Scandinavian ice sheet started after isotope stage 5e, slightly later than the growth of the North American or Antarctic ice sheets. During isotope stages 5d and 5b, "medium-sized" Scandinavian ice sheets developed. They disappeared nearly entirely during the Peräpohjola and Tärendö interstadials, which correlate with isotope stages 5c and 5a, respectively (Fig. 8).

At the beginning of isotope stage 4 (around 75 ka B.P.), an ice sheet grew that according to the model presented here (Fig. 8) persisted in the central areas of Scandinavia until the final deglaciation after 10 ka B.P. During the same time period the occurrence of considerable fluctuations along the western margin of the ice sheet has been demonstrated. Some thermoluminescence and U-series dates contradict the model presented and indicate that the ice also melted in the central areas sometime around 30–40 ka B.P. During the Torvastad, Bø, and Ålesund interstadials, here correlated with isotope stages 5a, mid 3, and late 3, respectively, the faunal assemblages indicate that a branch of the Atlantic Current swung into the Norwegian Sea.

References

Andersen, B. G., Sejrup, H. P., and Kirkhus, Ø. (1983). Eemian and Weichselian deposits at Bø on Karmøy, SW Norway: A preliminary report. *Norges Geologiske Undersøkelse* **380**, 189–201.

Andersen, S. T. (1961). Vegetation and its environment in Denmark in the early Weichselian Glacial. *Danmarks Geologiske Undersøgelse* **(II) 75**, 1–175.

——. (1965). Interglasialer og interstadialer i Danmarks kvartær. *Meddelelser Dansk Geologisk Forening* **15**, 486–506.

——. (1975). The Eemian freshwater deposit at Egernsund, South Jylland, and the Eemian landscape development in Denmark. *Danmarks Geologiske Undersøgelse Årbog 1974*, 49–70.

Averdick, F. R. (1967). Die vegetationsentwicklung des Eem-Interglazials und der Frühwürm-Interstadiale von Odderade/Schleswig-Holstein. *Fundamenta* **2**, 101–125.

Beaulieu, J. L., and Reille, M. (1984). A long Upper Pleistocene pollen record from Les Echets, near Lyon, France. *Boreas* **13**, 111–132.

Behre, K. E. (1989). Biostratigraphy of the last glacial period in Europe. *Quaternary Science Reviews* **8**, 25–44.

Behre, K. E., and Lade, U. (1986). Eine Folge von Eem und 4 Weichsel-Interstadialen in Oerel/Niedersachsen und ihre Vegetationsablauf. *Eiszeitalter und Gegenwart* **36**, 11–36.

Belanger, P. E. (1982). Paleo-oceanography of the Norwegian Sea during the past 130,000 years: Coccolithophorid and foraminiferal data. *Boreas* **11**, 29–36.

Bergersen, O. F., and Garnes, K. (1981). Weichselian in central South Norway: The Gudbrandsdal Interstadial and the following glaciation. *Boreas* **10**, 315–322.

Bowen, D. Q. (1978). "Quaternary Geology." Pergamon Press, Oxford.

Chaline, J., Mojski, J. E., and Meyer, K. D. (1980). Report on the symposium Vistulian stratigraphy, Poland 1979. *Boreas* **9**, 151.

Chappell, J., and Shackleton, N. J. (1986). Oxygen isotopes and sea level. *Nature* **324**, 137–140.

Dricot, E., Pettillon, M., and Seret, G. (in press). When and why did glaciers grow or melt in the Vosges Mountains, France? *Paleoklimatologie* **1**. Maintz.

Fægri, K. (1960). Maps of distribution of Norwegian plants. I. The coast plants. *Universitetet i Bergen Skrifter* **26**, 1–134.

Fulton, R. J. (1984). Summary: Quaternary Stratigraphy of Canada. *In* "Quaternary Stratigraphy of Canada – A Canadian Contribution to IGCP Project 24" (R. J. Fulton, Ed.), pp. 1–5. Paper 84–10, Geological Survey of Canada.

——. (1986). Quaternary stratigraphy of Canada. *In* "Quaternary Glaciations in the Northern Hemisphere" (V. Sibrava, D. Q. Bowen, and G. M. Richmond, Eds.) pp. 207–209. *Quaternary Science Reviews* **5**.

Grootes, P. M. (1978). Carbon-14 time scale extended: Comparison of chronologies. *Science* **200**, 11–15.

Grüger, E. (1979a). Die Seeablagerungen von Samerberg/Obb. und ihre Stellung im Jungpleistozän. *Eiszeitalter und Gegenwart* **29**, 23–34.

——. (1979b). Comment on "Grande Pile peat bog: A continuous pollen record for the last 140,000 years" by G. M. Woillard. *Quaternary Research* **12**, 152–153.

Harting, P. (1874). De bodem van het Eemdal. *Versl. Med. Kon. Akad. Wetensch.* **(II) 8**. (Not seen: as cited in Zagwijn, 1961)

Helle, M., Sønstegaard, E., Coope, R. G., and Rye, N. (1981). Early Weichselian peat at Brumunddal, southeastern Norway. *Boreas* **10**, 369–379.

Hirvas, H., Korpela, K., and Kujansuu, R. (1981). Weichselian in Finland before 15,000 B.P. *Boreas* **10**, 423–431.

Hirvas, H., and Kujansuu, R. (1981). Quaternary stratigraphy and chronology in northern Finland. *In* "Glacial Deposits and Glacial History in Eastern Fennoscandia" (G. I. Gorbunov, Ed.), pp. 5–25. Kola branch of the USSR Academy of Sciences, Apatity.

Hirvas, H., and Nenonen, K. (1987). The till stratigraphy of Finland. *Geological Survey of Finland Special Paper* **3**, 49–63.

Hütt, G., Punning, J. M., and Mangerud, J. (1983). Thermoluminescence dating of the Eemian-Early Weichselian sequence at Fjøsanger, western Norway. *Boreas* **12**, 227–231.

Jessen, K., and Milthers, V. (1928). Stratigraphical and palaeontological studies of interglacial freshwater deposits in Jutland and north-west Germany. *Danmarks Geologiske Undersøgelse* **(II) 48.**

Johnson, W. H. (1986). Stratigraphy and correlation of the glacial deposits of the Lake Michigan lobe prior to 14 ka BP. *Quaternary Science Reviews* **5**, 17–22.

Kellogg, T. B., Duplessy, J. C., and Shackleton, N. J. (1978). Planktonic foraminiferal and oxygen isotopic stratigraphy of Norwegian deep-sea cores. *Boreas* **7**, 61–73.

Korpela, K. (1969). Die Weichsel-Eiszeit und ihr Interstadial in Peräpohjola (nördlisches Nordfinland) im licht von submoränen sedimenten. *Annales Academia Scientiarum Fennicae* **(A) 99**, 1–109.

Kosack, B., and Lange, W. (1985). Das Eem-Vorkommen von Offenbüttel/Schnittlohe und die Ausbreitung des Eem-Meeres zwichen Nord- und Ostsee. *Geologishes Jahrbuch* **A86**, 3–17.

Kukla, G. J. (1977). Pleistocene land-sea correlations. *Earth-Science Reviews* **13**, 307–374.

Lagerbäck, R. (1988a). The Veiki moraines in northern Sweden—widespread evidence of an Early Weichselian deglaciation. *Boreas* **17**, 469–486.

——. (1988b). Periglacial phenomena in the wooded areas of Northern Sweden—relicts from the Tärendö Interstadial. *Boreas* **17**, 487–499.

Lagerbäck, R., and Robertsson, A. M. (1988). Kettle holes—stratigraphical archives for Weichselian geology and palaeoenvironment in northernmost Sweden. *Boreas* **17**, 439–468.

Larsen, E., Gulliksen, S., Lauritzen, S.-E., Lie, R., Løvlie, R., and Mangerud, J. (1987). Cave stratigraphy in western Norway; multiple Weichselian glaciations and interstadial vertebrate fauna. *Boreas* **16**, 267–292.

Liestøl, O. (1962). Areas and number of glaciers in Norway. *Norsk Polarinstitutt Skrifter* **114**, 32–54.

Lundqvist, J. (1967). Submoräna sediment i Jämtlands län. *Sveriges Geologiska Undersøkning* **Serie C 618**, 1–267.

——. (1971). The interglacial deposit at the Leveäniemi mine, Svappavaara, Swedish Lapland. *Sveriges Geologiska Undersøkning.* **Serie C 658**, 1–163.

——. (1981). Weichselian in Sweden before 15,000 B.P. *Boreas* **10**, 395–402.

——. (1986). Stratigraphy of the central area of the Scandinavian glaciation. *Quaternary Science Reviews* **5**, 251–268.

Makowska, A. (1982). Paleogeographic environment for Eemian marine transgressions on the lower Vistula. *Biuletyn Instytutu Geologicznego* **343**, 31–49.

Mangerud, J. (1970). Interglacial sediments at Fjøsanger, near Bergen, with the first Eemian pollen-spectra from Norway. *Norsk Geologisk Tidsskrift* **50**, 167–181.

——. (1977). Late Weichselian marine sediments containing shells, foraminifera and pollen at Ågotnes, western Norway. *Norsk Geologisk Tidsskrift* **57**, 23–54.

——. (1981). The Early and Middle Weichselian in Norway: A review. *Boreas* **10**, 381–393.

——. (in press). The Scandinavian ice sheet through the last interglacial/glacial cycle. *Paleoklimatologie* **1**. Maintz.

Mangerud, J., Andersen, S. T., Berglund, B. E., and Donner, J. J. (1974). Quaternary stratigraphy of Norden, a proposal for terminology and classification. *Boreas* **3**, 109–127.

Mangerud, J., Sønstegaard, E., and Sejrup, H. P. (1979). Correlation of Eemian (interglacial) Stage and the deep-sea oxygen isotope stratigraphy. *Nature* **277**, 189–192.

Mangerud, J., Sønstegaard, E., Sejrup, H. P., and Haldorsen, S. (1981). A continuous Eemian-Early Weichselian sequence containing pollen and marine fossils at Fjøsanger, western Norway. *Boreas* **10**, 137–208.

Martinsson, D. G., Pisias, N. G., Hays, J. D., Imbrie, J, Moore, T.C., Jr., and Shackleton, N. J. (1987). Age dating and the orbital theory of the ice ages: Development of a hig-resolution 0 to 300,000-year chronostratigraphy. *Quaternary Research* **27**, 1–29.

Menke, B. (1982). On the Eemian interglacial and the Weichselian glacial in northwestern Germany (vegetation, stratigraphy paleosols, sediments). *Quaternary Studies in Poland* **3**, 61–68.

Menke, B., and Tynni, R. (1984). Das Eeminterglacial und das Weichselfrühglazial von Rederstall/Ditmarschen und ihre Bedeutung für die mitteleuropäische Jungpleistozän-Gliederungen. *Geologisches Jahrbuch* **Reihe A, 76**, 3–120.

Miller, G. H., and Mangerud, J. (1986). Aminostratigraphy of European marine interglacial deposits. *Quaternary Science Reviews* **4**, 215–278.

Miller, G. H., Sejrup, H. P., Mangerud, J., and Andersen, B. G. (1983). Amino acid ratios in Quaternary molluscs and foraminifera from western Norway: Correlation, geochronology and paleotemperature estimates. *Boreas* **12**, 107–124.

Molfino, B., Heusser, L. H., and Woillard, G. M. (1984). Frequency components of a Grande Pile pollen record: Evidence of precessional orbital forcing. *In* "Milankovitch and Climate, Part 1" (A. Berger, J. Imbrie, J. Hays, G. Kukla, and B. Saltzman, Eds.), pp. 391–404. D. Reidel, Dordrecht.

Mojski, J. E. (1985). "Quaternary." Part 3b of "Geology of Poland." Publishing House Wydawnictwa Geologiczne, Warsaw.

Mörner, N. A. (1981). Weichselian chronostratigraphy and correlations. *Boreas* **10**, 463–470.

Müller, H. (1974). Pollenanalytische Untersuchungen und Jahresschichtenzälungen an der eemzeitlichen Kieselgur von Bispingen/Luhe. *Geologisches Jahrbuch* **A21**, 149–169.

Reusch, H. H. (1913). En notis om vore havdannede huler. *Norsk Geologisk Tidsskrift* **2**, 22–23.

Richmond, G. M., and Fullerton, D. S. (1986). Introduction to Quaternary glaciations in the United States of America. *Quaternary Science Reviews* **5**, 3–10.

Sejrup, H. P. (1985). Amino acid ratios: not only time and temperature. *In* "Abstracts 14th Arctic Workshop, Arctic Land-Sea Interactions," pp. 219–223. Bedford Institute of Oceanography, Dartmouth, Nova Scotia.

——. (1987). Molluscan and foraminiferal biostratigraphy of an Eemian-Early Weichselian section on Karmøy, southwestern Norway. *Boreas* **16**, 27–42.

Seret, G. (1967). "Les systèmes glaciaires du bassin de la Moselle et leurs einseignements." Societé Royal Belge de Géographie, 1–577.

Seret, G., and Roucoux-Woillard, G. (1978). The glaciations in the Vosges Lorraines. *In* "Führer zur Exkursionstagung des IGCP-Projektes 73/1/24 Quaternary Glaciations in the Northern Hemisphere" (B. Frenzel, Ed.), pp. 1–30. Deutsche Forschungsgemeinschaft, Bonn-Bad Godesberg.

Shackleton, N. J. (1969). The last interglacial in the marine and terrestrial records. *Proceedings Royal Society London*. **B. 174**, 135–154.

Sjørring, S. (1983a). The glacial history of Denmark. *In* "Glacial Deposits in North-West Europe" (J. Ehlers, Ed.), pp.163–179. A. A. Balkema, Rotterdam.

——. (1983b). Ristinge Klint. *In* "Glacial Deposits in North-West Europe" (J. Ehlers, Ed.), pp. 219–226. A. A. Balkema, Rotterdam.

St. Onge, D. A. (1987). The Sangamon Stage and the Laurentide Ice Sheet. *Géographie physique et Quaternaire* **41**, 189–198.

Suggate, R. P. (1974). When did the Last Interglacial End? *Quaternary Research* **4**, 246–252.

Turon, J. L. (1984). Direct land/sea correlations in the last interglacial complex. *Nature* **309**, 673–676.

Welten, M. (1981). Verdrängung und Vernichtung der anspruchsvollen Gehölze am Beginn der letzten Eiszeit und die Korrelation der Früwürm-Interstadiale in Mittel-und Nordeuropa. *Eiszeitalter und Gegenwart* **31**, 187–202.

——. (1982). Pollenanalytische Untersuchungen im Jüngeren Quartär des nördlichen Alpenvorlandes der Schweiz. *Beiträge zur Geologischen Karte der Schweiz* **156**, 1–174.

West, R. G. (1984). Interglacial, interstadial and oxygen isotope stages. *Dissertationes Botanicea* **72**, 345–357.

Woillard, G. M. (1975). Recherches palynologiques sur le Pleistocene dans l'est de la Belgique et dans les Vosges Lorraines. *Acta Geographica Lovaniensia* **14**, 1–118.

——. (1978). Grande Pile peat bog: A continuous pollen record for the last 140,000 years. *Quaternary Research* **9**, 1–21.

Woillard, G. M., and Mook, W. G. (1982). Carbon-14 dates at Grande Pile: Correlation of land and sea chronologies. *Science* **215**, 159–161.

Wright, H. E., Jr. (1977). Quaternary vegetation history—some comparisons between Europe and America. *Annual Reviews Earth Planetary Sciences* **5**, 123–158.

Zagwijn, W. H. (1961). Vegetation, climate and radiocarbon datings in the Late Pleistocene of The Netherlands. Part I: Eemian and Early Weichselian. *Mededelingen van de Geologische Stichting Nieuwe Serie 14,* 15–45.

——. (1975). Indeling van het Kwartair op grond van veranderingen in vegetatie en klimaat. *In* "Iȍlichting bij geologische overzichtskaarten van Nederland" (W. H. Zagwijn and C. J. van Staaldwinen, Ed.), pp. 109–114. Rijko Geologische Dienst. Haarlem.

——. (1983). Sea-level changes in The Netherlands during the Eemian. *Geologie en Mijnbouw* **62**, 437–450.

3
Peat Growth

RICHARD S. CLYMO

Peatlands

Lake and marine sediments, ice caps, and peat are the main repositories of detailed information about late Quaternary landscapes. Peat, with which this chapter is concerned, is widespread, locally abundant, and easily sampled. About three-quarters of all peatlands are in the USSR and Canada; a single peatland complex in western Siberia is about 1,800 × 800 km (Walter, 1977). Peatland inventories differ in their definitions and accuracy. One recent estimate is that 297 Mha is bog-covered and 210 Mha is swamp-covered—a total of about 3.4% of the Earth's land surface (Matthews and Fung, 1987). An earlier estimate (Kivinen *et al.*, 1979) gave 420 Mha. If one assumes a mean depth of 2 m and a mean dry-bulk density of 0.1 g cm^{-3} in bog-covered areas, then there is about 240 Gt of carbon in peat—Sjörs (1980) estimates 300 Gt. This last is about four times the amount of carbon fixed on the Earth's land surface in a year. In a suitable environment peat accumulates moderately but conveniently rapidly: a rule of thumb is a rate of about 100 cm in 1,000–2,000 years. The record it entombs is not confused by inwash or flow. The peat itself is predominantly the remains of plants that once grew on the surface but that have decayed incompletely—a fact that seems to have been first recognized by William King writing in 1685 (Gorham, 1953: 262): "your light spungy turf is nothing but a congeries of the threads of . . . moss . . . [which] is so quick growing a vegetable, that it mightily stops the springs, and contributes to thicken the scurf especially in red bogs." The peat thus retains both a macroscopic record of the plants that formed it and a local and regional pollen record. The record may extend back 5,000–9,000 years. The record of industrialization over the last 400 years in the form of metals, soot, fly-ash particles, and the like is also retained.

Pollen analyses at intervals as close as 1 mm are now being used to investigate patterns of activity that span one or a few years only (for example, Sturlodottir and Turner, 1985), and analyses of macroscopic remains are being used to elucidate details of ontogeny, both constructive (Janssens, 1983) and destructive (Tallis, 1985, 1987).

The rate of accumulation of peat varies greatly both in space and in time. Indeed the very idea of accumulation is widely misunderstood, as I explain later. The use of the record preserved in peat requires an understanding of the peat-accumulation process. What is accumulation? At what rate has peat accumulated? How are the small structures such as hummocks, lawns, hollows, and pools formed? How do they come to form a mosaic on the peatland surface? In what circumstances does peat-accumulating vegetation spread? Is there a limit to its spread? What part do productivity, decay, chemistry, hydrology, and ecology have in the accumulation process? Most of these questions require quantitative answers. Thirty-seven years ago Gorham, writing of the study of peatlands, recorded that "generally speaking, ecology has not yet advanced much beyond the stage of observation and description" (1953: 272). I hope to show here that we now have sufficient knowledge to begin to make quantitative predictions.

The Peat-Accumulation Process

Most peats accumulate as a consequence of waterlogging. Dead leaves, stems, and branches fall on the surface; roots grow down into the peat and die anachronistically; and mosses grow up and shade their own lower branches which then die *in situ*. Fungi, bacteria, invertebrate grazers, oxygen, and moisture allow aerobic decay, but at first the dead plant parts retain their macroscopic structure so that water can flow easily between the structural elements.

As decay weakens the dead plant matter, and as more dead matter (and perhaps a seasonal load of snow on top of unfrozen peat) is added to the surface, the structure collapses (Fig. 1). The dry-bulk density increases and, more important, the space between structural elements decreases. The hydraulic conduc-

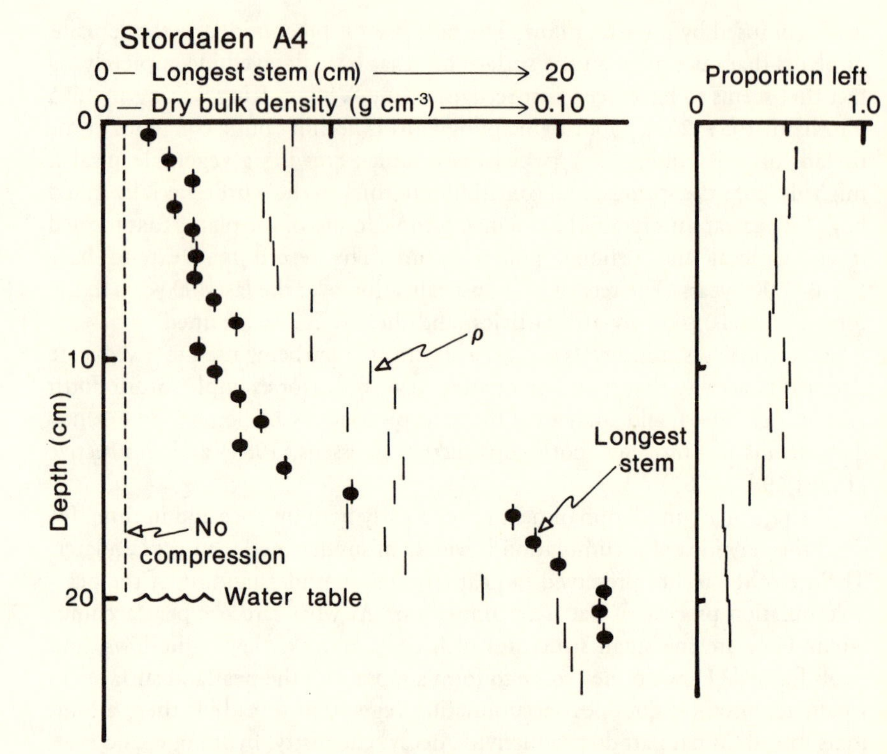

Fig. 1. Dry bulk density (ρ), length of the longest stem, and inferred decay (at right) in a core of *Sphagnum fuscum* peat from the Stordalen small mire, Abisko, northern Sweden, August 1976 (Clymo, unpublished data). The core was collected by means described by Clymo (1988). The proportion of the original dry mass left is calculated with the assumption that the extent of compaction can be inferred from the maximum length of *Sphagnum* stem in a slice. The maximum possible length was 22 cm—the width of the core—so the extent of decay below 19 cm is underestimated. See Fenton (1980) and Johnson *et al.*, (1990) for further details.

tivity decreases by several orders of magnitude. As long as precipitation and inflow (if any) exceed evapotranspiration, the water table will remain in the upper zone of high hydraulic conductivity, and water will seep away laterally. The water at the water table is oxygenated, and decay continues aerobically to a small distance below the water table. But oxygen can be replenished only by diffusion from above, and this is a slow process: the rate of diffusion of oxygen in water is barely 1/10,000th that of oxygen in air. Continuing aerobic decay therefore causes the peat to become anoxic and further decay is only by anaerobic processes, which (in peat) are much slower than aerobic ones. Consequently peat accumulates.

The accumulation of peat therefore involves an interaction between plant productivity, hydrology, and the processes of decay. A full account would require numerous qualifications. For example, in the Arctic and Antarctic, low temperature in permafrost may substitute for anoxia as a cause of low rates of decay.

The importance of the distinction between the surface zone of aerobic decay and the underlying anoxic zone of slow decay has been gradually recognized during the last 40 years or so, as experimental measurements have accumulated. (It was suspected much earlier than this.) In translation the Russian terms for the two layers are *active* and *inactive* or *inert* (Ivanov, 1953). These terms in English are confusing because *active* means something quite different to geologists interested in permafrost than it does to ecologists, while the lower zone, far from being inactive, has crucial effects on peatland morphology. To avoid these problems Ingram (1978) introduced the general term *acrotelm* for the largely oxygenated surface layer with high hydraulic conductivity and within which the water table fluctuates, and the term *catotelm* for the underlying, permanently saturated and mainly anoxic layer of low hydraulic conductivity.

Requirements and Systems for Peat Accumulation

There are many general accounts of the sorts of peat-accumulating systems (e.g., Gorham, 1957; Moore and Bellamy, 1974; Gore, ed., 1983) and innumerable regional and floristic accounts. Of particular historic or recent importance are the works of Weber (1908) in Germany; Cajander (1913) and Eurola *et al.* (1984) in Finland; Osvald (1923), Sjörs (1948), and Malmer (1962) in Sweden; Crampton (1911) and Tansley (1939) in Britain; Kulczýnski (1949) in Poland; Anderson (1983) in Malesia; and Rigg (1940, 1951), Conway (1949), and Heinselman (1970) in North America. The absence of any reference to Soviet work is unfortunate. The best account in English of one aspect of it—hydrological—is the translation by Thomson and Ingram of Ivanov (1975). In recent years some of the most exciting developments have been in the study of the enormous peatlands of eastern North America with their ovoid islands and water tracks. The work begins with Glaser *et al.* (1981), and Wright and Glaser (1983), and is partly summarized by Glaser (1987) and Damman and French (1987). Ideas about pool formation there have been elaborated by Foster *et al.* (1983, 1988) and Foster and Fritz (1987). This list is necessarily selective and personal. It may be supplemented by reference to the bibliographies made by Field and Goode (1981), and Gorham *et al.* (1985), and to the references in Gore, ed. (1983).

Most workers have recognized the interaction of water supply, water chemistry, and species-composition of the peat-forming vegetation. This is not the place for another synthesis but it is worth noting a few of the characteristics generally thought to be important.

The Chemistry of the Water supply

When the concentration of inorganic solutes is low, bog vegetation containing relatively few species develops. Shrubs of the heath family (Ericaceae), cotton-grasses (*Eriophorum* spp.), and especially bog-mosses (*Sphagnum* spp.) are ubiquitous. The last are particularly important because, though they can thrive only when the concentration of solutes is low, they are able to make the water around them acidic by a process of cation exchange (Clymo, 1963, 1984b); this and the low concentration of solutes make the environment unsuitable for most other plant species. Colonization by *Sphagnum* often marks profound ecological change.

The concentration of solutes may be low because the source of water is direct precipitation, or because the water has flowed over insoluble rocks, typically granite or sandstones. Where concentrations are too high for most species of *Sphagnum,* fen vegetation richer in species develops. Sedges, herbs, and trees, such as *Alnus* and *Salix,* are common.

The terms *ombrotrophic* (rainstorm-fed) and *minerotrophic* are used to describe vegetation nourished in these ways. Some authors would say that minerotrophic vegetation in parts of Finland, for example, may be more boglike than ombrotrophic vegetation in western Ireland (where *Cladium mariscus* and *Schoenus nigricans* grow in what is, by other criteria, bog). The concentration, nature, and rate of supply of solutes are the primary determinants; the source is secondary. Floristic distinctions may be subtle.

A Means for Keeping the Peat Waterlogged

Three principal mechanisms of waterlogging are recognized. First, water may percolate through rocks or soil and emerge as springs or lines of seepage or simply as upwelling over a large area. Retention in a porous medium evens the flow rate. If water collects from a catchment and is funneled into a lower area, the effect is as if the precipitation were greater. Water of this kind often has a relatively high concentration of solutes and supports species-rich fen vegetation. But if the rocks are resistant to weathering, the emerging water may allow *Sphagnum* to establish.

A second category of water supply is that in small, relatively deep lakes. Here a mat of vegetation may encroach on the water from the edge of the lake. The water is replenished from a catchment, and the surface of the vegetation mat falls and rises as droughts and rain dictate (Green and Pearson, 1968). The water table is always about the same distance below the surface of the vegetation mat and not far below it. It is possible for *Sphagnum* and associated plants to estab-

Table 1. Electrical conductivity G_{corr} (with the contribution from H^+ removed) and pH at various points in Cranberry Moss after rain in July 1987

	G_{corr} ($\mu S\ cm^{-1}$)	pH
Inflow stream	390	6.5
Peripheral fen	180	5.5
Open lake	240	6.6
Lake, below floating vegetation	250	6.5
Rain	25	5.3
Hollow in *Sphagnum* mat	35	4.5
Hummock in *Sphagnum* mat	38	4.8

lish or to persist on such floating mats even when the chemistry of the lake water is unsuitable, provided that there is sufficient rain to maintain a downward washing of the mat. At Cranberry Moss on the border between Staffordshire and Cheshire in the English Midlands, a lake about 100 m across and 15 m deep is three-quarters covered by a floating mat of *Sphagnum, Eriophorum, Vaccinium oxyoccos,* and *Empetrum* about 1.5 m thick. The lake is surrounded by a 10–20-m-wide wooded fen. Drainage from fertilized fields runs in a ditch around the fen and, at one point, through a side branch into the lake. The ditch frequently floods and overflows into the fen and lake. The chemical conditions are shown in Table 1. The insulation of the surface of the floating *Sphagnum* mat from the lake water is remarkable. Areas of this kind are interesting but are usually small and of much less quantitative importance than the third kind.

The third mechanism occurs where the peat has accumulated above the regional groundwater table. Waterlogging in such circumstances implies a dynamically maintained, domed water table and precipitation as the source of water. Precipitation must exceed evaporation in most weeks in the year, but the proportion of rain days (days with 1 mm or more of precipitation) is important too. These conditions give rise to domed, raised bogs or, in constantly humid climates, to blanket bogs, both with abundant *Sphagnum*. It is astonishing that the first clear explanation in the English ecological literature of the cause of the domed water table seems to be that of Ingram (1982), though the idea had occurred to others including Weber (1908), Granlund (1932), Wickman (1951), and Ivanov (1953). Ingram pointed out that although the water table after rain would sink quickly through the acrotelm, because the hydraulic conductivity is high, it would sink very much more slowly through the catotelm. The plant structures have collapsed, the space between elements is much smaller, and the hydraulic conductivity, which is approximately related to the fourth power of the size of spaces between elements, is correspondingly and dramatically reduced. Water therefore seeps very slowly through the catotelm. The next rain rapidly raises the water table again into the more porous acrotelm. Most ecologists—citations would be invidious—had vaguely invoked capillary forces, though Granlund (1932) had pointed out that the height to which water could rise experimentally in peat was barely 0.5 m. This implies a radius of curvature of the meniscus of about 30 μm and spaces between elements of 0.03–0.06 mm. We know that the spaces within plant cell walls are much smaller than this, but even if they were continuous, the *rate* at which water could ascend would be determined by the fourth power law and would be inadequate to counter evaporation. Grandlund's 0.5 m is the practical limit. The same conclusion could have been reached by noting that the water table was close below or, in pools, above the bog surface and realizing that in a hole in jelly there is no water table. If the capillary explanation were adequate, a hole a handspan across in a raised bog should show a water table near the base of the bog and close to the regional water table, not close to the bog surface.

Changes in Time

The classical sequence of changes shows a shallow lake becoming filled by reed swamp, followed by fen vegetation, and this in turn replaced by bog. The whole process has been called terrestrialization. In other cases it seems that impeded drainage has allowed bog or fen to spread directly over mineral soil, often replacing forest—a process called paludification. A particularly impressive array of evidence—buried soil profiles, tree stumps, peat stratification, diatoms, and ^{14}C dates—has been assembled for the Kräckebäcken mires in Sweden (Foster and Fritz, 1987). Most boreal peatlands have probably originated by paludification (Sjörs, 1961).

It is apparent that a particular peatland may contain peat formed under different influences and to various extents at different times in the past. Some of the best examples are those now being revealed in North America. The most important difference between the peatlands of the Glacial Lake Agassiz and those studied in Europe seems to be that in the former the underlying calcareous till is included in the water-circulation patterns. A computer simulation (Siegel, 1983) suggested that recharge on a raised bog with a water table only 30 cm above the regional water table might cause water flow as deep as 15 m into the underlying till. It is salutary to discover that measurements in the field (Almendinger *et al.*, 1986; Siegel and Glaser, 1987) suggest that at some times of year water movement is upward into raised bogs in these mire complexes as well as into spring-fen and water track, and that in the *Sphagnum*-covered, raised bog, chemical conditions are unsuitable for *Sphagnum* as little as 50–100 cm below the surface, recalling those in the floating mat of Cranberry Moss (Table 1). That this delicate balance has been upset in the past is shown in places where fen vegetation has replaced bog (Glaser, 1987).

Surface Patterns

Many boreal peatlands show conspicuous surface patterns composed of the microforms hummock, lawn, hollow, and pool; Sjörs (1961) offers color photographs. The boundaries are rather arbitrary for, as Bragg (1982) showed, the frequency of surface altitudes on a regular grid shows no evidence of distinct categories. Ecological boundaries may be seen, however: the lower limit of *Calluna vulgaris* on hummocks in Europe is an example. These microforms may be organized into microtopes such as the ridge-and-flark structure (elongate hummocks and intervening pools) and a series of microtopes may form a bog mesotope. Large bog complexes are macrotopes. The arrangement of microtopes may be orderly as, for example, in the north of Scotland where Bragg and Ingram (reported in Ingram, 1987) found a series from the bog center outward of the six microtopes: perennial (large) pools, ridge-flark, hummock-hollow, dwarf sedge, dwarf shrub, and fen. This regularity has been widely recognized in the USSR (Ivanov, 1957) and seems to be found in eastern North America as well (Glaser and Janssens, 1986).

There has been much speculation about the origin of pools on raised bogs and blanket bogs, beginning with Crampton (1911:53), who believed that "they can only be due to movements in the peat." This idea was developed by Pearsall (1956), who suggested wrinkling or tearing of the surface as causes. Tearing is the likely explanation in a few cases where a peat mass is flowing very slowly over an underlying step in the rock, as it does at Muckle Moss (Pearson, 1960) and at Coom Rigg (Chapman, 1964), both in the north of England. But this mechanism is probably rare. Consensus seems to be growing that large pools on bogs develop where there is least flow of water and that they develop from hollows in which either the rate of production is lower than it is in their surroundings, or the rate of decay is higher, or both (Boatman, 1983; Hulme, 1986; Ingram, 1987; Foster *et al.*, 1988). But there is still much room for debate, and the extent of pool development differs enormously from site to site (Fig. 2). The origin of strings and flarks is in much the same state (Foster *et al.*, 1983; Foster and Fritz, 1987): there are interesting hypotheses and accumulating evidence.

For the immediate purpose it is sufficient to recognize that these surface features are widespread, though not ubiquitous. Can they be found below the surface, too? Sernander (in von Post and Sernander, 1910) suggested that in places where a mosaic of active hummocks and hollows (the "regeneration complex") exists, there might be a cyclic alternation in time of hummock and hollow leaving regular lenticular structures in peat. The evidence was slight, and the structures may have been inferred from the surface features, but the idea was accepted and used by Osvald (1923) to interpret his mainly phytosociological work on Komosse. The hypothesis was spread by Clements (1916) and by Tansley (1939), who published an account of cores from two bogs in Ireland, collected

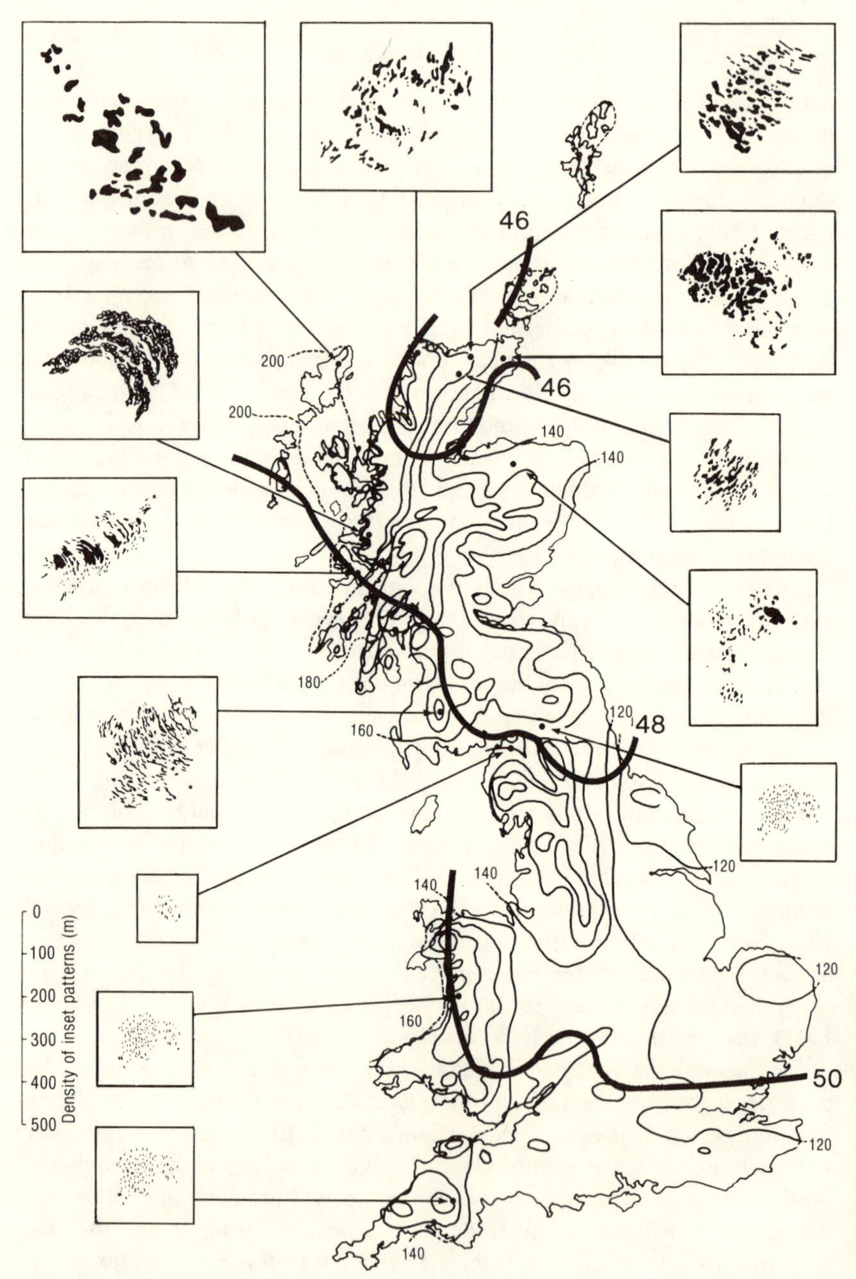

Fig. 2. The distribution of patterns of pools on bogs in Britain. The fine lines are isopleths of mean wet day yr^{-1}. The broad lines are isotherms (°F) of mean daily mean temperature (46, 48, 50 °F = 7.8, 8.9, 10.0 °C). From Lindsay *et al.* (1988). (With permission: Nature Conservancy Council.)

in company with and interpreted for him by Osvald. The idea was given even wider circulation by Watt (1947) in his influential Presidential Address to the British Ecological Society. This was its apogee. Some ecologists were already uneasy about it. First Walker and Walker (1961), then Casparie (1969), Barber (1981), and Svensson (1988) sought in vain for evidence on cut peat faces. Backéus, who also sought without success for unpublished evidence in Sernander's notebooks, comments that all the detailed work has "taken place in maritime areas where Osvald himself did not expect cyclic regeneration to play a major role" (1987:122). Yet Tansley writes, "In the summer of 1935 the author . . . [examined] . . . some Irish raised bogs in the company of Professor Osvald who interpreted the details of their vegetation and peat structure in the field. . . . A boring was made by Professor Osvald . . . in the middle of the typical 'regeneration complex.' . . . A detailed record is annexed (Table XXI)" (Tansley, 1939: 688 and 690). This appears to be the only detailed evidence interpreted as supporting the hypothesis. There are few areas more maritime than Ireland.

Evidence of such structures in the lower layers of peat is more dubious. Cut peat faces in these zones are less often exposed for detailed examination, and selective decay may have removed the evidence. We cannot, at present, be sure whether or not strong surface patterning is a relatively recent development.

What has emerged very clearly from the detailed work on cut peat faces is abundant evidence of subsurface structures at least as complex as those we see on the surface today. These imply bewildering local variations in space and time of plant productivity and decay. The variations are not just local. There are often "recurrence" horizons (Granlund, 1932) in peat, which can be traced laterally over a whole bog, where a layer of highly humified peat is sharply delimited from an overlying layer of less-humified peat. These horizons usually are interpreted as a response to changes in climate but are often not contemporary in bogs in the same district. There seems to be an element specific to the internal dynamics of the growth of the bog. An additional complication is that bogs at different altitudes, with different temperature and hydrologic regimes, may respond differently to the same change in regional climate if one bog is thereby moved over a response threshold while the other is not (Conway, 1948).

A particularly well documented example is Store Mosse near Värnamo in South Sweden (Svensson, 1988). Macrofossils were recorded in 2,019 samples from 49 cores, 30 on a single 1,800-m transect. Three *Sphagnum* phases were observed. Basal peats of many types of fen were replaced over large areas about 7,000 yr B.P. at a depth 4–4.5 m below the present surface by *Sphagnum* peat—mostly by *S. fuscum*-type but in some places by *S. rubellum*-type. This peat becomes more humified with remains of Ericaceae and *Eriophorum vaginatum*. The second *Sphagnum* phase, which begins about 2,400 yr B.P. at 3 m below the present surface, is more varied with *S. cuspidata*-type replaced by *S. fuscum*- or *S. rubellum*-types. Again a highly humified layer occurs, then the third *Sphagnum* phase begins about 1,000–1,200 yr B.P. at 2 m below the surface with *S. cuspidata*-type replaced by *S. magellanicum*-type, but with *S. rubellum*-, *S. fuscum*-, or *S. imbricatum*-types below present-day strings. The highly humified layers correspond with periods of low lake levels, and the beginning of the second and third *Sphagnum* phases with high lake levels (but the *S. fuscum* stage began when lake levels were low). It seems clear from this enormous mass of evidence that the changes were synchronous over a large area of the bog and most are probably attributable to change in climate, though it is not known why different species should have been favored.

We must conclude that no single hypothesis is likely to explain more than a small proportion of the observed and observable phenomena of peat growth. But we can examine the processes that contribute to peat growth, measure their rates, and try to show what consequences must follow. We should not expect to find real cases that fit our predictions exactly—peatlands are too complicated—but we may hope to find the trends we predict, in simple cases at least. It is to this that I now turn. Most of what follows is concerned with bogs because they are simpler and better known. A brief examination of surface hydrology precedes consideration of decay processes.

Hydrology of the Acrotelm

The range of conditions in the acrotelm may be understood by considering the fluctuations in the water table at the surface of a bog. A comprehensive review is given by Ingram (1983). The water table responds rapidly to rainfall. The hydraulic conductivity of the acrotelm (Fig. 3) has been measured by an ingenious field method (Bragg, 1982). Conductivity decreases close to exponentially down the acrotelm, which thus behaves like a V-notch weir, as can be seen in the hydrograph shown in Fig. 4. Where the conductivity is high, the water table falls quickly. Where it is lower, it falls in steps, probably reflecting the importance of evapotranspiration during the day. A long-term summary as a residence curve is shown in Fig. 5. For half the time the water table is within a few centimeters of the mean level, but the curve is asymmetric. In droughts the water table sinks 20–30 cm. The layer between the extreme lows and the mean water table is thus occasionally oxygenated and this, one assumes, affects the rate of decay and the nature of the microorganisms. Those in the layer that is alternately oxygenated and anoxic may be facultatively anaerobic.

Decay

The immediate agents of decay—fungi and bacteria—are present in peat (Table 2). Fungal mycelia are abundant in the oxygenated zone but become much rarer

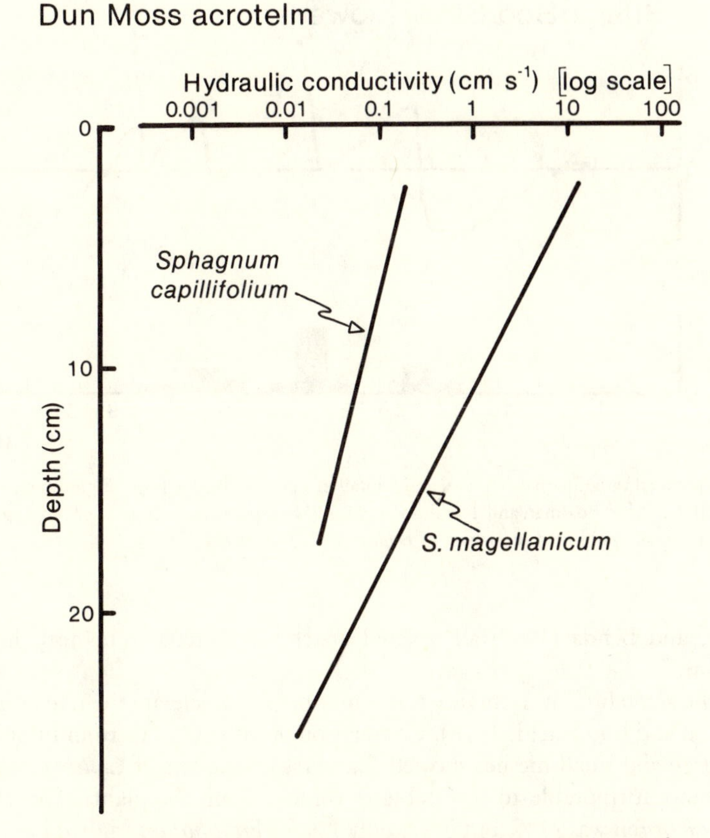

Fig. 3. Variation of hydraulic conductivity with depth in two low hummocks on Dun Moss, in the southern Grampian foothills of Scotland. The lines shown were obtained by differentiating a line, assumed to be straight, of field measurements in relation to depth (Bragg, 1982). The r^2 for the field data was > 0.99. (With permission from author.)

as one moves down into the anoxic zone. The same is true of aerobic bacteria. Anaerobic bacteria are as abundant as aerobic ones in the topmost zone—they may survive in microanoxic pockets there. In the lower layers they do not seem to become more abundant, but they do not diminish. Burgeff (1961) was unable to isolate viable bacteria from deep in the catotelm of raised bogs, and there is a persistent though implausible belief that because *Sphagnum* can be used as a wound dressing, it and peat formed from it are sterile. Anyone who has tapped a pocket of methane deep in the peat while peat-boring may suspect that there is some microbial activity; Waksman and Stevens (1929), Waksman and Purvis

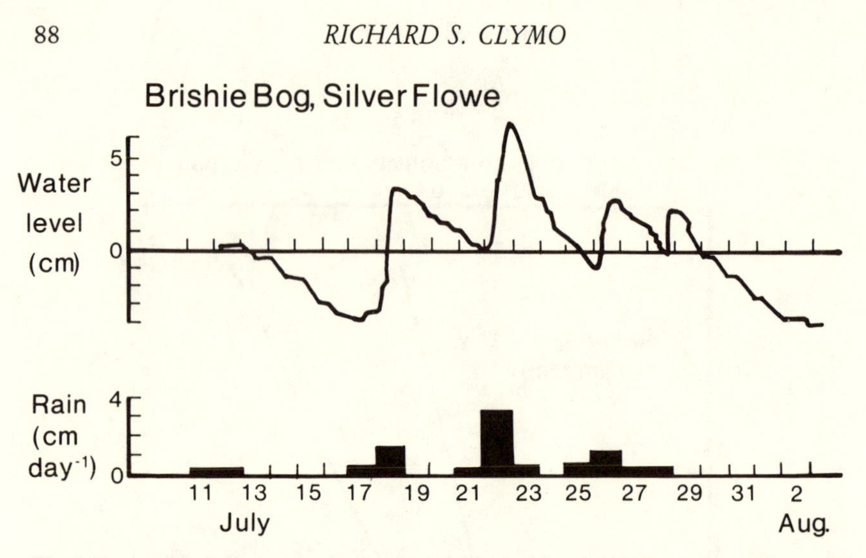

Fig. 4. Temporal variation in water level and rainfall in a pool on Brishie Bog, Wigtonshire, southwest Scotland. After Boatman and Tomlinson (1973). (With permission: *Journal of Biogeography* 10:223; originally published in *Journal of Ecology* 69 [1981]:897–918.)

(1932), and Benda (1957) all isolated anaerobic bacteria from deep in the catotelm.

In the acrotelm, invertebrate grazers may help to accelerate the rate of decay (Coulson and Butterfield, 1978). Comparison of losses of mass from litter bags with large and small meshes showed that 43% of the loss of *Calluna vulgaris* shoots was attributable to invertebrates comminuting the plants. For *Rubus chamaemorus* it was 27%, but it was only 1% for *Eriophorum vaginatum* leaves and even less for *Sphagnum recurvum*. Here is the first sign of selectivity among

Table 2. Abundance of fungi and bacteria in the surface of blanket-bog peat at Moor House, northern England

Depth (cm)	Zone	Fungi: stained mycelium ($m\ g^{-1}$)	Aerobic bacteria ($10^5\ g^{-1}$) (range)	Anaerobic bacteria ($10^5\ g^{-1}$) (range)
0–5	"Litter"	2450	9–260	9–250
5–12	Dark brown	1030	6–150	32–200
12–20	Green-brown	750	11–76	16–500
20–32	Red-brown	200	1–42	28–260

Note: Fungal abundance was estimated by the length of stainable mycelium; bacterial potential (of many different groups) was estimated by counting colonies on nutrient agars of various sorts (Collins *et al.*, 1978). Water-table depth was about 15 cm. The red-brown zone was probably anoxic.

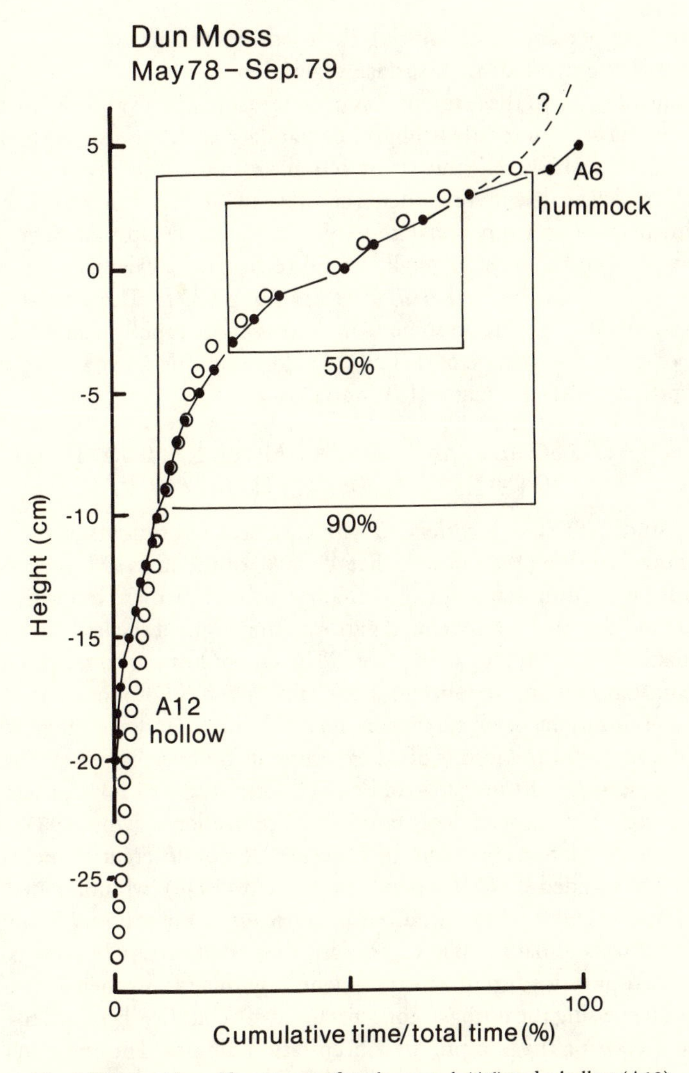

Fig. 5. Water-table residence curves for a hummock (A6) and a hollow (A12) on Dun Moss, Scotland. The vertical axis is centered on 50% residence. The small rectangle encloses 25–75% residence; the large, 10–90% residence. Measurements were made at approximately weekly intervals so occasional brief, high-water levels will have gone unrecorded. These are indicated by the broken line. From Bragg (1982, with permission from author).

species—an important factor limiting the usefulness of macrofossils in peat as indicators of the previous mire-surface vegetation.

The rate of decay in the acrotelm has been measured by a variety of methods. The loss in mass, or in tensile strength, of standard cellulose strips is useful for comparing sites; however, the strips tell little about the decay of natural materials, which are losing constituents of different kinds at different rates as the acrotelm, defined as a functional zone, gradually moves up past them. Litter bags have proved the most generally useful technique. Reviews of this sort of work have been made by Heal *et al.* (1981), among others. Three questions are frequently asked. (1) Why do some plants decay more rapidly than others? (2) How does decay rate change over time? (3) How does the rate of decay change with depth at which the material is found?

Why Do Some Plants Decay More Rapidly Than Others in the Same Habitat?

In many studies the rate is correlated with chemical constituents of the plants. For example, Ohlson (1987) found that 45–80% of the mass of leaves of *Carex rostrata* from a spring fen and from an adjacent intermediate fen were lost in one year, and that the rate depended partly on the origin of the leaves and partly on the habitat in which they were placed. The rate of loss was correlated in various circumstances with the concentrations of nitrogen, phosphorus, and potassium. Concentrations of all three were relatively high, and the range was not great. In a survey of the results of 22 experiments on bog plants, in which the nitrogen concentrations were low and ranged from 0.5% to 2.1%, there was a positive correlation between decay rate and concentration (Clymo, 1983). There are multiple correlations between the concentration of different elements. This problem was avoided by Coulson and Butterfield (1978), who obtained plant material with increased concentrations of nitrogen or phosphorus by applying specific fertilizers to natural bog vegetation, then used material harvested from the fertilized plots for litter-bag experiments. By this means they were able to show that increasing the nitrogen concentration, specifically, increased the decay rate, but increasing the phosphorus concentration did not. The graph in Clymo (1983) can thus be treated as a regression. The loss of mass is 10% yr^{-1} at 0.7% nitrogen concentration, and 50% at 1.7%, giving a slope of 40. No doubt other intrinsic factors operate too, but chemistry seems to be of great importance.

How Does Decay Rate Change over Time?

The usual technique is to put out litter bags and to recover them at intervals for several years. In practice it is rare to be able to make measurements for more than five years. The results of Heal *et al.* (1978) were analyzed by Clymo (1984a), who showed that for *Rubus chamaemorus, Calluna vulgaris,* and *Eriophorum vaginatum* it was not possible to distinguish between the hypotheses (1) that the rate of loss was a constant proportion of the original mass in each year for five years or (2) that it was a constant proportion of what was left at any time. The problem arises because these data have a fairly large coefficient of variation, and the two hypotheses do not become readily separable until 70–80% of the original material has disappeared, by which time other sorts of bias—collapse of the litter bags, loss of fragments, and so on—become important. It is also the case that the longer an experiment continues the more likely it is that the effects of age become confounded with changes in the environment as the acrotelm moves up past the litter bags.

How Does the Rate of Decay Change with Depth?

Again the most frequent technique is to put litter bags at known depths. The results differ in detail (Clymo, 1965; Heal and French, 1974) but agree in showing that the rate of decay in the acrotelm is relatively high and that it declines in the catotelm (Fig. 6). The plant material placed at different depths was from a single batch, so the lower rate of decay in the catotelm must be a consequence of the different environment. It is not simply that readily decomposable material has been removed and only refractory material remains.

As the acrotelm moves up past a particular piece of organic matter, the rate of decay of the plant material changes, partly because its chemical composition changes (because of the accumulated effects of decay thus far) and partly because the thermal, hydraulic, chemical, and microbiological environment changes too. What we need to know ultimately is how the rate of decay changes as a consequence of all these factors. It may take 20–100 years for the acrotelm to pass by, and research grants do not last that long. In a few favorable cases there are indirect ways of making such measurements. The earliest, and still the best measurements, were those by Baker (1972) on the antarctic moss *Chorisodontium aciphyllum,* which, with other species of moss, forms peat banks 1–2 m deep. Annual increments of about 2 mm yr^{-1} can be identified. Baker assumed that the rate of addition of dry matter had been approximately constant and by weighing segments spanning a known age was able to calculate the rate of decay (Fig. 9, discussed below). A similar approach was used by Fenton (1980) on the same moss banks and by Johnson *et al.* (1990) on *Sphagnum* in southern Sweden. Fenton measured the angle that the moss stems make as they collapse and compared the dry-bulk density calculated from this with that measured. He assumed that differences were attributable to decay. Johnson measured stem lengths in slices of known thickness for the same purpose. It is surprising that all three found indications that decay in the acrotelm was slow near the surface, then became rapid, then slowed again. Figure 1 (apart from the top 1-cm slice)

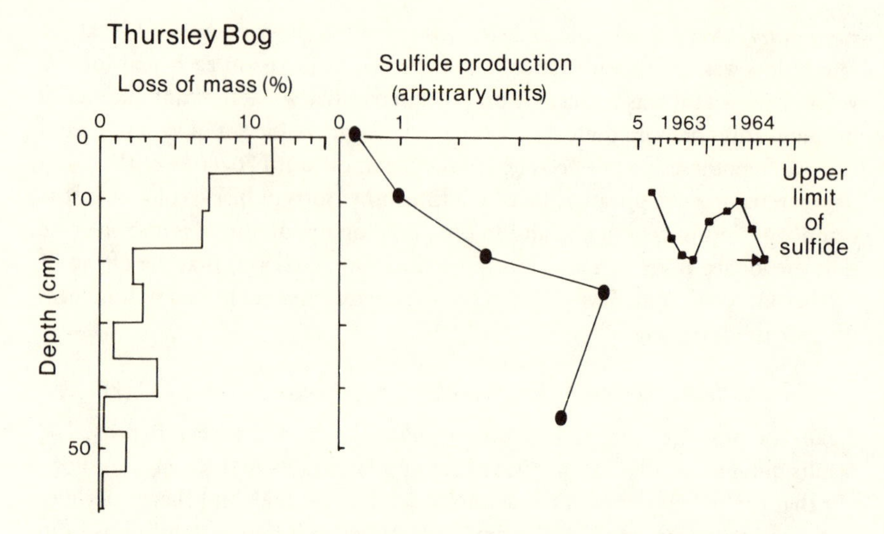

Fig. 6. Profiles in the surface of Thursley Bog, southern England, of decay rate of *Sphagnum papillosum* in litter bags, sulfide production in enrichment cultures, and seasonal upper limit of sulfide detectable on a silver wire in the field. From Clymo (1965). (With permission: *Journal of Ecology* 53: 753.)

shows the same effect. Extension of this work is hindered by lack of reliable techniques for dating samples in the acrotelm.

Before leaving the acrotelm it is worth considering the question: is the rate of decay in hollows markedly different from that in hummocks and ridges? The question is important because large pools, once formed, seem to be long-lived if not permanent (e.g., Foster *et al.*, 1988). Many contain little vegetation, and it is easy to see why they persist. In one plausible scheme a pool begins as a hollow in which productivity is less than in its surroundings, decay is more rapid, or both. It then accumulates peat less rapidly and so falls behind. But the intrinsic rate of loss of mass from bog litter put into hollows was only about half that of similar material put into hummocks (Farrish and Grigal, 1985). Claricoates (unpublished) has measured the efflux of carbon dioxide and methane throughout the year from two bogs in northern England. The mean fluxes of $CO_2 - C$ ($n = 30$) from hollow, lawn, and hummock sites from which green matter had been removed were 80, 61, and 320 mg m^{-2} day^{-1} and of $CH_4 - C$ were 15, 10, and 1 mg m^{-2} day^{-1}. Some of these gas effluxes would be from the catotelm, but the strong seasonal variation implies that most efflux is from the acrotelm. The loss from hummocks is about five times that from hollows. Silvola (1986) measured much-increased rates of carbon-dioxide efflux from a Finnish bog in which the water table had been lowered. Higher rates of CO_2 efflux from hummocks would be explained if the acrotelm there were thicker than in hollows. Certainly the water table is farther below the surface of hummocks than it is below that of hollows, though in some cases fluctuations below hollows are greater than below hummocks (Bragg, 1982).

Nils Malmer and Bo Wallén of Lund offer an interesting ecological explanation for this. Hollows commonly contain *Sphagnum* and sedges. The sedges are able to flourish in waterlogged conditions, which kill ericaceous shrubs, but the sedges lack woody stems and give little support to *Sphagnum*. By contrast, in drier conditions on hummocks, woody ericaceous shrubs are able to grow and suppress sedges. The stems provide support around which *Sphagnum* grows up. This in turn stimulates the development on newly buried stems of adventitious roots, and renders the plants potentially immortal (Forrest and Smith, 1975; Wallén, 1980). The woody structures thus maintain a thicker acrotelm than do the sedges in wet hollows, and this in turn allows a longer time for decay. Even if the productivity of the drier sites were greater, the flux of peat from the acrotelm down into the catotelm might be less than in wetter sites. The essential and striking element of this hypothesis is that the hummocks, which are higher, are adding peat to the catotelm more slowly than are the hollows, which are lower. As the hollow slowly rises relative to the hummock, so the thickness of the hummock acrotelm decreases, and it begins to add peat to the catotelm more rapidly. There is thus a negative-feedback mechanism that prevents a hummock drawing away from a hollow but also prevents it being overtaken by the hollow. Hummocks and hollows should tend to persist, though their relative importance will be affected by climate wetness, as appears to be the case (Walker and Walker, 1961; Barber, 1981). There is an important distinction here between hollows, which have a complete vegetation cover, and pools, which do not. Pools are failed hollows.

Decay in the catotelm has not been studied widely. The rates are low and therefore not easily measured, and because they are low, they were thought to be unimportant. One obvious bacterial activity is the production of sulfide which, at the pH prevailing in bogs, is mostly in the easily detectable form H_2S. The upper limit of detectable blackening on a silver wire or on a silver-plated sheet has often been used to record the depth at which the peat has become effectively anoxic. An example of the fluctuations is shown in Fig. 6. It seems that sulfide production is at its greatest at the top of the catotelm about 20–40 cm below the surface (Fig. 7). There is a good correlation between sulfide activity and "redox potential" (Claricoates, unpublished). The amount of sulfide produced need not necessitate much decay. Direct measurements of methane efflux do imply decay, however, and have been recorded by, among others, Clymo and Reddaway (1971), Svensson (1980), Svensson and Rosswall (1984), Harriss *et al.* (1985), and Claricoates (1990). Efflux of methane from wetter sites is usually greater by a factor of 2–8, as already listed, than from sites with a lower

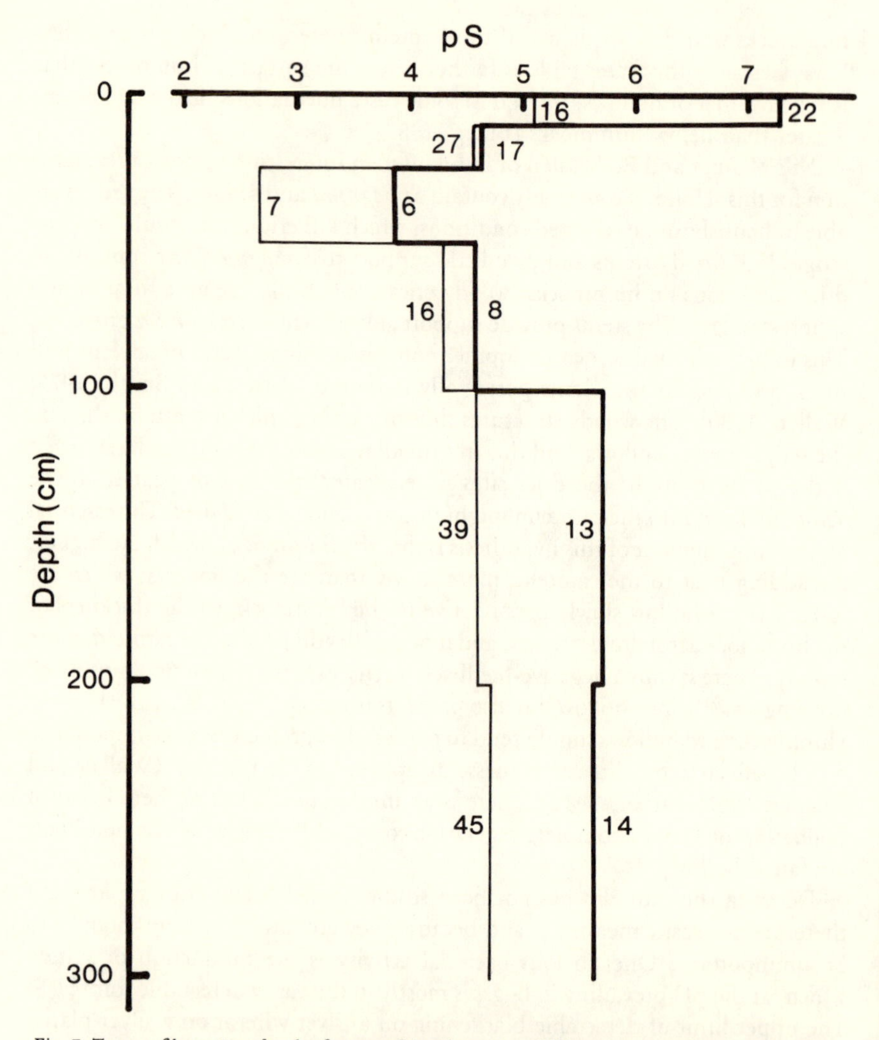

Fig. 7. Two profiles to 3 m depth of mean pS (analogous to pH) in peat at two sites at Moor House, northern England. Samples were brought to the surface and immediately placed in antioxidant buffer at high pH, thus trapping all forms of sulfide. Vertical lines plot average pS of the number of samples shown in each interval. Claricoates (1990). (With permission from author.)

water table. Erratic high values—perhaps the escape of bubbles—seem common. It is not obvious why there should be such differences if most of the methane has been produced in, and diffused from, the catotelm. It is possible that aerobic, methane-oxidizing bacteria remove gas as it passes up through a hummock. Svensson and Rosswall (1984) show that methane-producing potential is

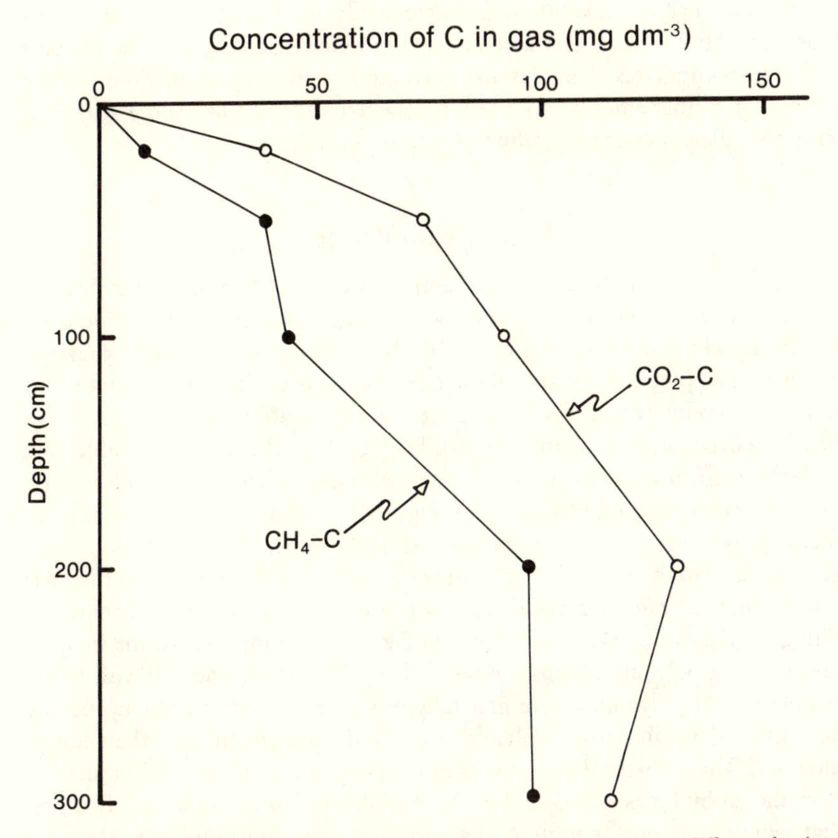

Fig. 8. Concentration of carbon from methane and carbon dioxide in peat at different depths at Moor House, northern England. Claricoates (1990). (With permission from author.)

greatest at the top of the catotelm (where pS and redox potential show minima), but they also found that methane concentration increased steadily down to 30cm—the greatest depth sampled. Claricoates (1990) has sampled peat at much greater depths and finds that the concentrations of both carbon dioxide and methane increase with depth (Fig. 8). The only plausible explanation of such profiles is that gases are still being produced in the catotelm, and are not simply fossil. Calculations based on the assumption that gases leave solely by diffusion are consistent with a rate of decay of about 0.0002 yr^{-1} (Clymo, 1984a).

The efflux values for CH_4 – C vary widely: Median values of about 20–50 mg m^{-2} day^{-1} seem common, but Harriss *et al.* (1985) found a mean of 114 mg m^{-2} day^{-1}. If this continued throughout the year it would amount to 42 g m^{-2} yr^{-1}—a significant fraction of surface productivity on bog sites. The explanation is probably that the measurements were made in August, in which month Claricoates (1990) found methane-efflux rates were at their highest, and perhaps 2–5 times winter rates. This seasonal fluctuation again suggests that most of the efflux is of methane newly produced in an active zone just below the acrotelm, with a smaller, nonseasonal efflux of catotelm methane.

Peat Accumulation

We can now consider how in a very simple case a peat bog might accumulate peat. If we start from nothing, the earliest stages of accumulation would be something like those shown in Fig. 9(b). To construct this one must know, or surmise, how productivity and decay rate change over time. To illustrate, assume that productivity is constant (year-to-year fluctuations, which certainly occur, have little effect on accumulation). For peat bogs there are no reliable data of decay as a function of age, confounded with depth, and so on, so let us use the Antarctic *Chorisodontium* data in Fig. 9. Two approaches are possible: either to make some ecologically credible, though oversimple, assumption, or to get a good fit to these data by some ad hoc function. As an example of the first, assume that the rate of decay is directly proportional to the mass of material. This gives the exponential decay curve in Fig. 9(a). For the second, the temptation to use a polynomial must be resisted: nearly always, one will wish to extrapolate, and polynomials are generally unstable and unrealistic beyond the data used to obtain them. Polynomials include straight lines as their lowest member. The unhelpful consequences of using them for peat accumulation were shown by Jones and Gore (1978). For illustration I have used a function that swings from one asymptote to another—a "wall function." This does not go through the origin, but in practical cases is close to it—within 2% for *Chorisodontium*. It tends toward a lower asymptote, implying that part of the plant matter is totally refractory. In the very long term this may be thought to be unrealistic, but in practice would not be invoked because by that time the material would be in the catotelm, which must be considered separately.

The accumulation of peat with these assumptions is shown in Fig. 9(b). Although the two curves in Fig. 9(a) differ substantially, there is little difference in the accumulation curves over the span of these data (about 40 years). Only at 60 years do they begin to diverge greatly. Thereafter the differences increase steadily because the exponential is tending toward an asymptote at p/α with a slope of zero, while the wall function tends toward an upward slope of pv (because the wholly refractory matter of proportion v simply accumulates).

The important point is that at some stage the acrotelm, defined by the predominantly aerobic decay processes, will have passed upward and left the peat as it now is in the quite different environment of the catotelm. The transi-

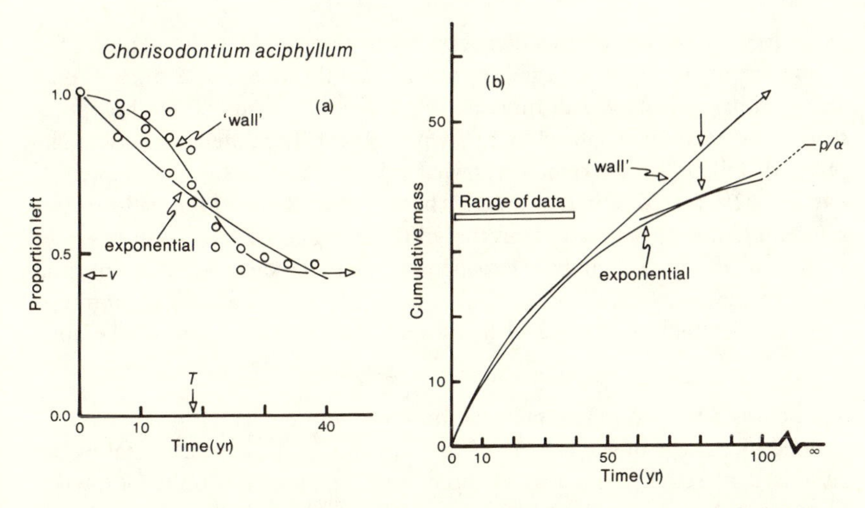

Fig. 9. (a) Decay of *Chorisodontium aciphyllum* in the maritime Antarctic. Redrawn from Baker (1972, with permission: *British Antarctic Survey Bulletin* 27: 126). The hollow curve is the best fit exponential, including 0, 1.0 as a fixed point; the sigmoid curve is the best fit of an ad hoc "wall function." (b) Cumulative mass of peat assuming constant rate of addition and decay according to the exponential or "wall function" of (a). The vertical arrows at 80 years indicate where the acrotelm has finally passed by. The slope at these points is the apparent rate at which the catotelm receives matter at its top, compared with the slope at the origin which is the rate at which it enters the acrotelm by plant production. If m_t is the mass at time t, then the exponential in (a) is $m_t/m_o = \exp(-\alpha t)$ where α controls the tightness of curvature. The "wall function" is:

$$\frac{m_t}{m_o} = 1 - \frac{(1 - v)}{(1 + \exp[-r(t - T)])}$$

where v is the lower bound, T is the time of steepest descent, and r controls the curvature. The best-fit values to these data are: $\alpha = 0.0215\ \text{yr}^{-1}$, $v = 0.434$, $T = 18.0$ yr, $r = 0.216$. The curves in (b) are obtained from $M_t = \int_0^t p\, m_t/m_o\, dt$ where p is the productivity (assumed = 1.0 for illustration). For the exponential $M_t = (p/\alpha)(1 - \exp[-\alpha t])$. For the wall function:

$$M_t = p\,\{tv - [(1 - v)/r]\,[\ln(1 + \exp[r(t - T)]) - \ln(1 + \exp[rT])]\}$$

tion will not be sharp: it may last for tens of years, and the conditions in the top of the catotelm may be sufficiently different (as Fig. 7 suggests) to justify recognizing an intermediate zone. But the difference between the acrotelm and all of the catotelm is so great that for exploratory purposes we may consider just two layers, and a sharp transition between them.

In Fig. 9(b) the transition is shown at 80 years, which for the *Chorisodontium* peat is about the age at which it is submerged by the top of the permafrost zone at 20 cm depth (Fenton, 1980). Two points of view must be considered.

The acrotelm, having reached this thickness, is now in a steady state. As it ascends, matter enters it at the top surface at a flux p (assumed 1.0 in Fig. 9), and leaves it at the bottom at a flux given by the slope of the accumulation curve at 80 years in Fig. 9(b). For the exponential case this rate is 0.18 units of mass area^{-1} time^{-1}. For the wall function it is $v = 0.43$ flux units. For bog peats, real values are in the range 0.1–0.2 (Clymo, 1984a). The difference between input and "fallout" to the catotelm is what is lost by decay while the acrotelm is passing: 82% and 57% in the two cases here. The thickness of the wall-function acrotelm is only 19% greater than that of the exponential one, but it passes on 43% of what it receives and the exponential one passes on only 18%. The exact shape of the decay curve in Fig. 9(a) is therefore of considerable importance. Once the acrotelm is established its *accumulation* rate becomes zero because:

$$\text{input} = \text{decay} + \text{fallout}$$

There is a temptation to take the mass (on an area basis) in the acrotelm divided by the age of the bottom of the acrotelm, and to speak of that as "accumulation rate." For Fig. 9(b), that gives 0.56 and 0.48 flux units for the wall function and exponential cases. These are actually the mean accumulation rate *while the acrotelm was first becoming established.* Once established the acrotelm no longer accumulates: The catotelm is the true site of accumulation of peat.

From the point of view of the catotelm, the fallout from the acrotelm given by the slope of Fig. 9(b) is influx. This material has been modified extensively in quality and quantity from what was originally produced. Some of the surprising consequences of having mixtures of materials with different decay characteristics are explored by Clymo (1984a). If the flux into the acrotelm is constant, then so is that into the catotelm. In general the influx to the catotelm is a smoothed version of that to the acrotelm. The catotelm is dark, water-saturated, anoxic, and nearly isothermal. The contrast with the acrotelm must be similar to that experienced by a secondhand-car salesman entering a monastery.

We know little about the rate of decay in the acrotelm. About the catotelm we know nothing except that on the evidence of Fig. 8, decay does continue. We must make assumptions, see where they lead, and try when possible to compare their predictions with reality.

The simplest realistic assumptions are of constant input to the catotelm and of decay directly proportional to the mass of material accumulated. (These are the same as one set of assumptions made for the acrotelm, but the parameter values are smaller in practice.) The consequences are shown in Fig. 10. There is an asympototic maximum depth at p'/α' (90 g cm^{-2}, equivalent to 9 m depth) compared with unlimited depth if there is no decay. Indeed, the depth is unlimited if there is any totally refractory material—the equivalent of $v > 0.0$ in the wall-function description of decay in the acrotelm. The true rate of accumulation is given by the slope of the convex curve at any time. Clearly it is p' and greatest at the start, and falls steadily toward zero as time passes. At all times

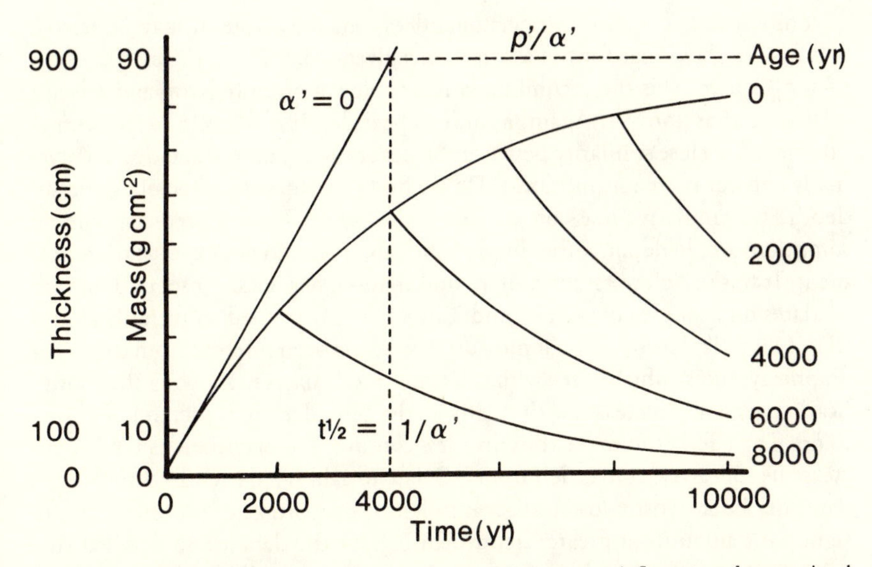

Fig. 10. Cumulative mass of peat in the catotelm assuming constant influx, p', and proportional rate of decay, α', with values 225 g m^{-2} yr^{-1} and 0.00025 yr^{-1} respectively. The depth scale assumes that the dry-bulk density is 0.1 g cm^{-3}. The solid straight line is of slope p' and shows what would happen if there were no decay. The horizontal broken line is the asymptotic limiting thickness at p'/α'. The descending concave curves show the position of parcels of peat starting at the surface at 2,000-yr intervals and sinking because of decay below them. The formal description is that $dM'/dt' = p' - \alpha' M'$ where M' is the accumulated mass in the catotelm at time (in the catotelm) t'. From this $M' = (p'/\alpha')(1 - \exp[-\alpha' t'])$.

the surface of the acrotelm would be green and the influx to it would be constant, producing in turn constant influx to the catotelm. Yet the true rate of peat accumulation in the catotelm decreases steadily, and *must* do so, whatever the shape of the decay curve, provided that there is some decay and that there is no totally refractory material. The first is almost certainly true; the second is plausible. The intuitive belief that a green, healthy surface must mean rapid accumulation is difficult to uproot. Even more difficult to dislodge is the belief that accumulation rate in the past can be measured by taking the mass or depth between two horizons and dividing by the difference in age of those horizons. The danger of this practice may be seen in Fig. 10 (right side). In this case we know that the influx to the catotelm has been constant throughout, yet the application of the usual procedure would show a steadily increasing "rate of accumulation" and would no doubt invoke hand-waving reference to internal peat-bog dynamics and to an increasingly moist climate. If there is any decay then such procedures and such "explanations" must be suspect.

The pattern at the right of Fig. 10 shows one way in which the assumptions of constant influx to, and proportional decay in, the catotelm may be tested. It is easy to show that for the assumptions given, then $M' = (p'/\alpha')(1 - \exp[-\alpha' t'])$ where M' is the accumulated mass below a given horizon and t' is age relative to that horizon. Again, a limit to peat depth at $M' = p'/\alpha'$ is a consequence. (The close similarity between the peat growth and the age-depth curves holds only for these assumptions.) Do we find the concave profile of age versus depth (as cumulative mass on an area basis), which Fig. 10 predicts? The assumptions are, in detail, rather implausible, so we should not expect close agreement. It was therefore astonishing to find that the best existing set of data with 55 dates on a single core (Aaby and Tauber, 1975) seemed to fit fairly closely (Fig. 11). Bulk density was not measured, so it has been inferred from the stratigraphic symbols, which agree with colorimetric estimates in showing that humification generally increases with depth, as dry bulk density is therefore inferred to do. The effect is to make the curve less concave. Other situations tend to increase its concavity: conversion from ^{14}C date to calibrated (dendrochronologic) date; dry bulk density lower at greater depth; proportion of ash increasing to significant amounts at greater depths. In Fig. 11 the data are so detailed that the temporary effects visible as recurrence surfaces are easily seen. These are no more than stumbles on a stately march through the millennia consistent with constant influx and constant proportional decay. The best-fit values are $p' = 0.005$ (s.e. $= 0.001$) g cm^{-2} yr^{-1}, and $\alpha' = 1.2 \times 10^{-4}$ (s.e. $= 0.9 \times 10^{-4}$) yr^{-1}. This influx of 50 g m^{-2} yr^{-1} is perhaps 10–20% of that to the acrotelm of present bog surfaces.

The same methods applied in other cases yield a mixed bag. Some are shown in Clymo (1984a), one in Smith and Clymo (1984), and others in Fig. 12. Of the 16 cases, 9 show concave curves, 4 appear straight, 1 is slightly convex, and 2 have more than one age "reversal," which, if omitted, could make them concave. Of the 5 straight and convex cases, 2 are based on depth as distance and might become more convex were their dry-bulk density known.

On balance we ought perhaps to picture the catotelm as having a limit to its depth: a limit at which the influx from the acrotelm is balanced by losses of carbon dioxide and methane from the whole of the catotelm. In any case calculation of "accumulation rate" cannot be made safely by the traditional procedure of dividing the depth or mass found today between two horizons by the difference in age of those horizons. The work of Tallis (1985) is the only case where any attempt at all has been made to allow for decay. The results were enlightening.

The Shape and Size of Raised Bogs

That precipitation has something to do with the growth and size of raised bogs has long been believed (Granlund, 1932; Wickman, 1951; Aartolahti, 1965).

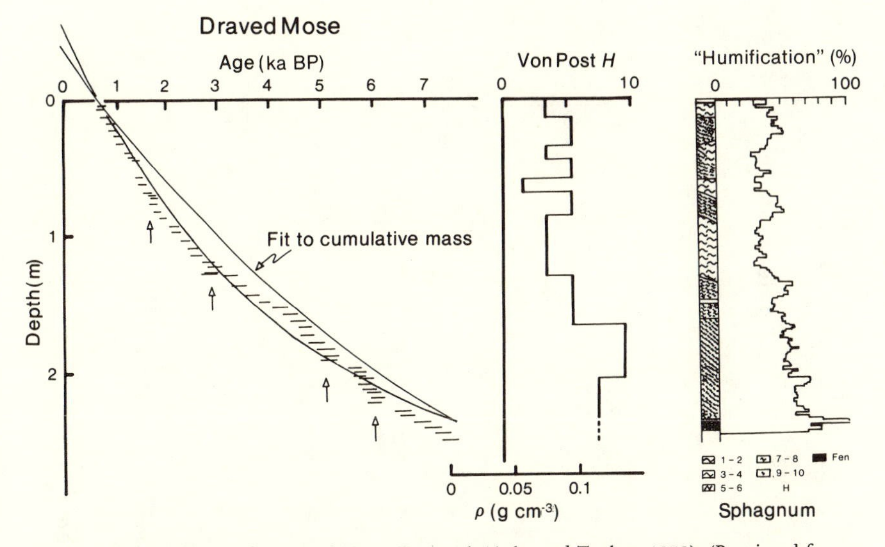

Fig. 11. Age versus depth for Draved Mose, Denmark (Aaby and Tauber, 1975). (Reprinted from Aaby and Tauber, 1975, from *Boreas* by permission of Universitets forlaget AS [Norwegian University Press].) The horizontal bars are calibrated dates with counting errors. The lower concave curve is the line of best fit to the depth-age curve using depth in linear measure. We need depth as cumulative mass (on an area basis). This was not measured but can be estimated from the humification symbols using $\rho' = 0.1\,H + 0.04$ g cm^{-3} where ρ' = dry-bulk density and H = von Post humification (Clymo, 1983, 1984a). Stratigraphy and a colorimetric measure of humification are at the right. To the left of them are the inferred H and ρ' values. These were used to get another line of best fit shown above the first one and arranged to coincide with the first at the youngest and oldest points. Overall mean $\rho' = 0.10$ g cm^{-3}. Vertical arrows mark "recurrence zones" where a period of slow peat growth was followed by a period of rapid growth. The surface of the peat has been cut away.

But it was only recently that Ingram (1982, 1983) pointed out that the low hydraulic conductivity of the catotelm is sufficient to account for a groundwater mound—a phenomenon well known to soil physicists, for example, Childs (1969). During periods of ample rainfall, excess water seeps laterally through the acrotelm. During droughts, however, the water table sinks into the catotelm as the catotelm begins to drain. It is probably the rare, long droughts that are critical: Ingram suggests that irreversible changes in the catotelm peat, which are known to occur on drying, may be important. The survival of the surface vegetation in such cases seems to be of equal importance. Ingram's hypothesis is that the shape of the raised bog in vertical section is determined by its hydrology, and that the acrotelm is constrained to conform to a shape determined in that way. For the simplest case of a homogeneous, isotropic catotelm, the groundwater mound should be close to a hemi-ellipse in cross section. For a raised bog

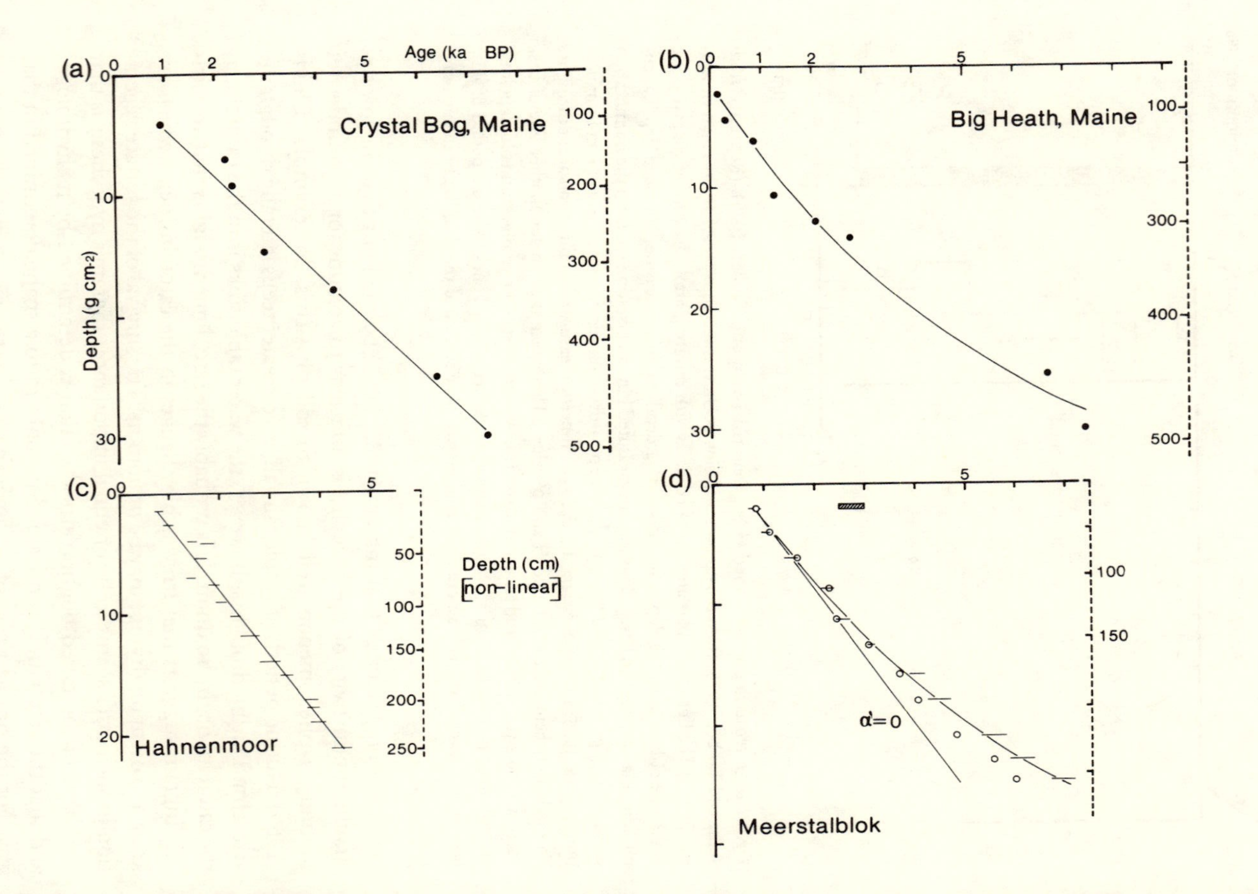
(a)
Age (ka BP)
Crystal Bog, Maine
Depth (g cm-2)
(b)
Big Heath, Maine
(c)
Depth (cm)
[non-linear]
Hahnenmoor
(d)
α=0
Meerstalblok

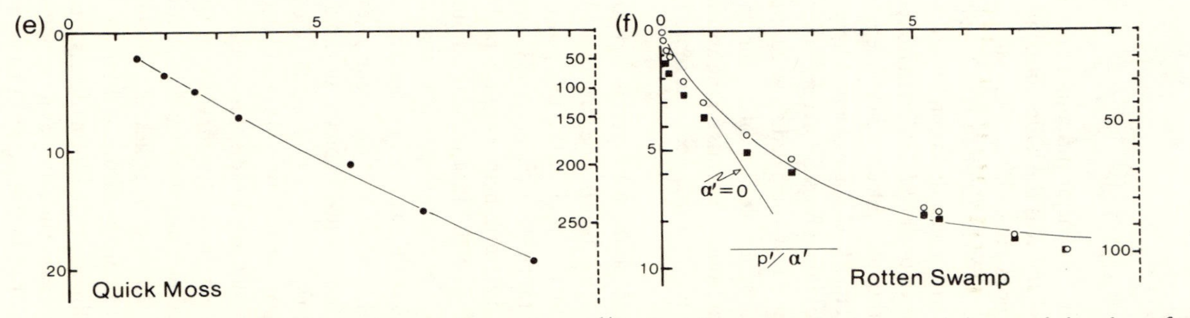

Fig. 12. Age versus depth in six bogs. In (a)-(e) the age is from calibrated ^{14}C. The left axes show depth as cumulative mass below the surface on an area basis. The broken axes at the right show the depth in linear measure at particular points, but these axes are not a simple transformation of those at the left. The model to which lines were fitted is described in the text.

(a) Crystal Bog, Maine (Tolonen *et al.*, 1983). (With permission: *Maine Geological Survey Bulletin* 33: 60.)

(b) Big Heath, Maine (Tolonen *et al.*, 1983). (With permission: *Maine Geological Survey Bulletin* 33: 60.)

(c) Hahnenmoor, Germany (Middeldorp, 1984). The bars show the counting-error limits. Middeldorp gives another less complete but strongly concave example for Engbertdijksveen, Netherlands. (With permission from the author, whose thesis concerned the variability rather than the regularity of the age/depth curve.)

(d) Meerstalblok K2, Netherlands (Dupont, 1985, and personal communication). Bars as for *c*. The open circles are the uncalibrated ^{14}C dates; calibration makes the curves more concave. The straight line shows what would be expected if there were no decay. (With permission from the author.)

(e) Quick Moss, northern England (Rowell and Turner, 1985, with permission: *Journal of Ecology* 73: 19). The dry-bulk density was calculated from the reported colorimetric estimate of humification (h) as $\rho' = 0.0012\ h\ (g\ cm^{-3})$. A plot using linear depth was markedly concave.

(f) Rotten Swamp, ACT, Australia (Clark, 1986, and personal communication, with permission). Dates were obtained from cumulative pollen density, scaled by a single calibrated ^{14}C date at 86 cm depth. Open circles are cumulative mass; filled squares are depth in linear measure but scaled to coincide with the mass scale at the oldest sample to show how the curve is made less concave using mass.

which is elliptical in plan view, with major and minor radii X and Y, then the height, h, above the regional water table is contained in:

$$\frac{U^*}{K} = h^2 \frac{(1/X^2 + 1/Y^2)}{(1 - x^2/X^2 - y^2/Y^2)}$$

where x and y are distances from the center along the major and minor axes, U^* is the net recharge in times of critical drought, and K is the hydraulic conductivity. The cross sections of two small, raised bogs in Scotland, Dun Moss and Ellergower Moss, are reasonably close to hemi-elliptical (Ingram, 1982, 1987), and the surface of Ellergower Moss gives a fit with mean deviation of 0.6 m to the equation above (Clymo, unpublished). The value of U^*/K inferred from the shape of both is about 0.001, which is close to the value calculated from separate measurements of U^* and K at Dun Moss (Ingram, 1982). This hemi-ellipse accords with early observations that the margins of raised bogs are steeper than the bog expanse—indeed the margins are recognized by the Swedish word *rand*. The margins of raised bogs, being the most accessible, are often the most damaged. The simple groundwater-mound description depends on the assumption that water flow is approximately horizontal, and this becomes more unrealistic at the margins as well.

The hemi-elliptical profile is a consequence of specific assumptions, in particular that the substratum below the peat is impermeable. In cases where the hydraulic conductivity of the substratum is similar to that of the peat, as it is thought to be in the Glacial Lake Agassiz area, for example (Siegel, 1983; Siegel and Glaser, 1987), one should not expect to find a simple hemi-elliptical profile.

Can a hemi-elliptical profile be combined with the idea of an asymptotic maximum depth, discussed in the previous section? It can, provided that one concedes that the rate at which plant mass enters the catotelm is constant only at the bog center. In the case analyzed by Clymo (1984a), the bog begins growth at a single focus on a flat, impermeable plain and gradually spreads out, maintaining the hydrological hemi-ellipse at all times. The consequences are shown in Fig. 13. One curious feature of the interaction between hydrology and decay is that the rate of growth at the edge of the bog increases steadily for a long time before decreasing again. To the unwary, ignoring decay, this effect seems even more marked (Fig. 13, left side). One must assume that the concomitant, complicated changes in the range at which the acrotelm supplies the catotelm (Clymo, 1984a) are mediated by changes in the depth or nature of the acrotelm forced by the underlying hydrological hemi-ellipse. If pool formation is a random process, so that the density of pools or perhaps their area as a proportion of the bog surface is a reflection of the length of time that a particular part of the bog has been in existence, then the density should show a pattern like that in Fig. 13b.

This sort of calculation is useful only in suggesting consequences—sometimes

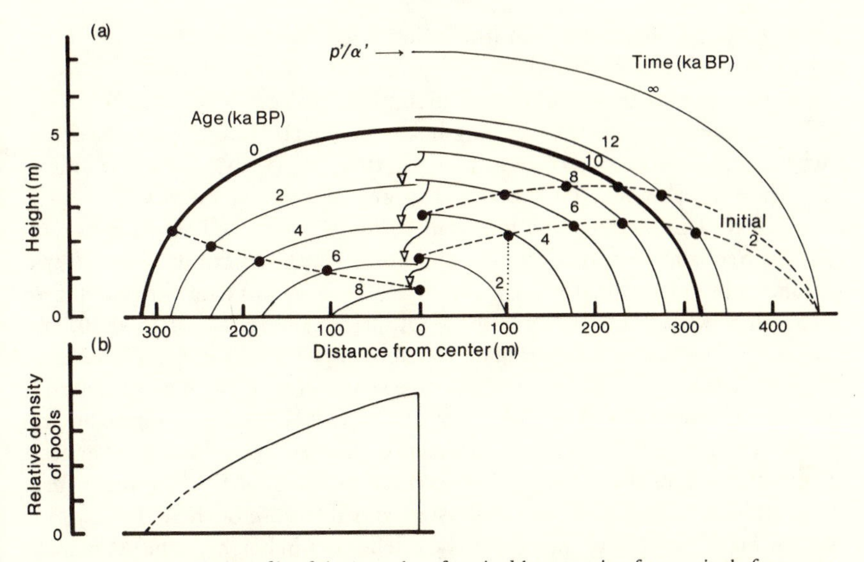

Fig. 13. (a) Hypothetical profile of the catotelm of a raised bog growing from a single focus, constrained by the hydrological groundwater mound in droughts and by $(p'/\alpha')(1 - \exp[-\alpha' t'])$ at the center. The vertical scale is 20 times the horizontal one. The right half shows the surface at 2,000-year intervals to 12,000 yr, and the final steady state. The left half shows for the specific age of 10,000 yr the position of earlier surfaces. Broken lines on the right show the amount of growth in the first 2,000 and 4,000 yr at the bog margin as it spreads. On the left, the broken line shows the apparent growth in the first 2,000 yr at the margin if decay is ignored. (b) Relative density of pools if pool formation is a random process in space and time.

unexpected—of apparently simple and reasonable assumptions. There can be few places where bog growth starts at a single focus and spreads. One possible example is Hammarmossen, which Granlund (1932) concluded had begun growth simultaneously over its whole area but which recent work (Foster *et al.*, 1988) with ^{14}C dating indicates spread from a single focus.

What happens when growth begins at the same time over a large area, or at several or many foci that then fuse? Some indication may be obtained from the work of Granlund (1932), who recorded the length (Swedish *langd*) and maximum depth of 792 Swedish bogs. Granlund categorized the bogs by district and within district by annual precipitation. To each subset he drew a freehand "limiting" curve showing that as size increased so did depth but in a convex manner. Wickman (1951) in an influential but rather confusing article pointed out that these "limiting" curves could be considered a single family, which is a close fit to $H^2/L = C$, where H is the maximum depth of peat, L is the "length," and C is a constant. (The hemi-ellipse, however, requires that $H^2/L^2 = C$.) Wickman managed to imply that the maximum depth is determined by precipitation, and this intrinsically plausible conclusion has been widely, if uncritically, accepted. What Granlund's "limiting" curves actually show is that there seems to be a relation between H and L but they set no maximum to H *unless* L is fixed.

If we return to the groundwater mound, it may be written as $U^*/K = aH^n/L^m$, where L is now the half length for a parallel sided bog with end drains (in which case $a = 1$) or L is the radius of a circular bog (with $a = 2$). The indices $m = n = 2$. This may be rearranged to give: $\log H = (m/n) \log L + \log (U^*/Ka)/n$. This straight line form can be tested against all Granlund's data (not his freehand "limiting" curves) and also against 115 Finnish bogs in two regional subsets (Aartolahti, 1965). There are 14 subsets of data altogether if we ignore three with 12 or fewer members. Of these, 13 have $r^2 = 0.55$–0.90, median 0.72. The values for m/n fall into two groups. For 9 subsets in Småland, southwest Häme, and north Satakunta, the values of m/n (with one exception) are in the range 0.48–0.56. For Svealand, Västerbotten, and Jämtland the values, with one exception, are in the range 0.71–0.80. It is clear that in general the relationship of these data is close to constant H^2/L or H^4/L^3, but there is no relation to precipitation or to bog size, and the cause of these groupings is not obvious. There seems therefore to be a deficiency in height, or rather in maximum depth, of peat if one takes the lateral spread of the raised bog as given and expects a hemi-ellipse. The obvious explanation is that most of these are bogs whose lateral extent was established early in their development, probably by topography, and whose profile has not yet become hemi-elliptical.

Will it ever become hemi-elliptical? Two possibilities may be recognized. In the first the size of the bog is such that the height of the hydrological ellipse is less than the decay limit, that is, $H < p'/(\alpha'\rho')$ (where ρ' is the mean dry-bulk density in the catotelm). In this case the outer zones are the first to approach the hydrological hemi-ellipse appropriate to L, and growth there slows down. The boundary between limited growth and the central area, which is still overwet, gradually contracts until the whole profile is hemi-elliptical. If the bog cannot spread it will just grow upward with the same profile until the $p'/(\alpha'\rho')$ limit is reached.

The second possibility is that the bog is so large that the hydrological hemi-ellipse cannot be completed before the $p'/(\alpha'\rho')$ limit is reached. For example, in lower Saxony the mean width of 64 raised bogs was 6,000 m (Eggelsmann, 1976). If a U^*/K value of 0.001 applied to them, as it does to the two bogs in Scotland, then $H = 95$ m. For $p' = 50$ g m^{-2} yr^{-1}, $\alpha' = 10^{-4}$, and $\rho' = 0.1$ g cm^{-3}—values close to those reported by Clymo (1984a)—then the $p'/(\alpha'\rho')$ limit would be 5 m. This is indeed close to the mean found by Eggelsmann, and much less than the hydrological limit of 95 m.

The center of these bogs is therefore much flatter than the hemi-ellipse appropriate to their width, and one assumes that they will never reach the hemi-elliptical profile. One may speculate (Fig. 14) that such bogs will also be wetter away from the margins, because the water table will not be driven down so far

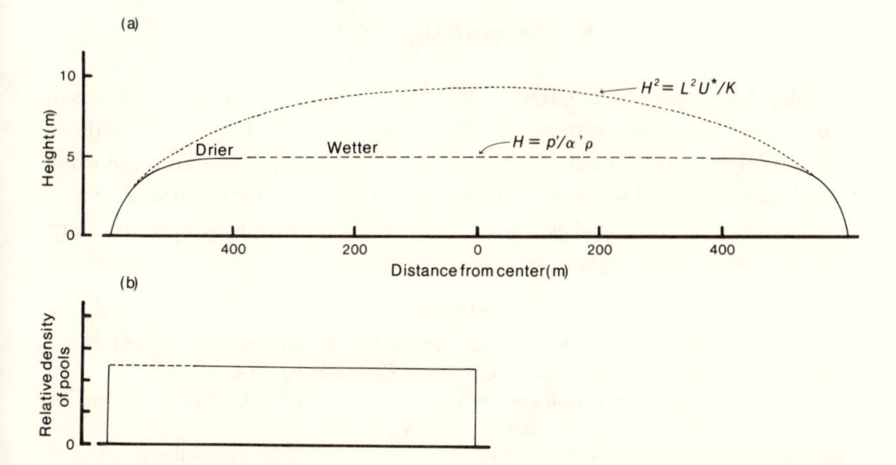

Fig. 14. (a) Hypothetical development of a raised bog of relatively large size. The dotted line is the hydrological hemi-ellipse for $U^*/K = 0.001$, $L = 300$ m. The decay limit at $p'/(\alpha'\rho')$ is for $p' = 50 \text{ g m}^{-2} \text{ yr}^{-1}$, $\alpha' = 0.001 \text{ yr}^{-1}$ and $\rho' = 0.1 \text{ g cm}^{-3}$. This depth is less than the hydrological one so the postulated limit is shown by the solid-and-dashed line, which approaches the hydrological limit at the margins but then falls well below it. (b) Relative density of pools if pool formation is a random process in space and time.

in droughts as it is in bogs that have reached the hemi-ellipse. The density of pools generated randomly in space and time would be nearly uniform across the surface (Fig. 14b), but hydrology might require a greater density of pools in the center. The gradient to drive water is less than it would be in a hemi-ellipse and may be compensated by the low resistance to acrotelm flow in pools.

In summary, U^* may be primarily determined by the requirements for plant survival in rare long droughts, and K may be primarily determined by the sort of peat. The size and profile of raised bogs may depend rather little on climate. If they are small (< 500 m across) then, it is likely that their profile will by now approximate the hydrological hemi-ellipse. But if they are larger and began growth over their whole area at once, then although their margins may be approaching the hydrological hemi-elliptical profile, their centers will not have done so because there has been insufficient time or because they have reached the decay limit, $p'/(\alpha'\rho')$. Such centers would be wetter than the margins.

We may thus glimpse how the outlines of a more exact understanding of peat growth are taking shape. But it is wise for those of us at work today to recall that, if we think we see farther than our predecessors, then so too do dwarfs on the shoulders of giants.

Acknowledgments

I thank J. Claricoates, R. L. Clark, L. M. Dupont, and L. Johnson for permission to use as yet unpublished results; O. M. Bragg, H. A. P. Ingram, N. Malmer, and B. Wallén, from whom I know that I have borrowed ideas; all those whose ideas I have absorbed without realizing or remembering their source; P. Ratnesar for continuing technical assistance; and the H. E. Wright Symposium Committee for the stimulus to write this chapter.

References

Aaby, B., and Tauber, H. (1975). Rates of peat formation in relation to humification and local environment, as shown by studies of a raised bog in Denmark. *Boreas* **4**, 1–17.

Aartolahti, T. (1965). Oberflächenformen von Hochmooren und ihre Entwicklung in Südwest-Häme und Nord-Satakunta. *Fennia* **93** (1), 1–268.

Almendinger, J. C., Almendinger, J. E., and Glaser, P. H. (1986). Topographic fluctuations across a spring fen and raised bog in the Lost River Peatland, northern Minnesota. *Journal of Ecology* **74**, 393–401.

Anderson, J. A. R. (1983). The tropical peat swamps of western Malesia. *In* "Ecosystems of the World Vol. 4B. Mires: Swamp, Bog, Fen and Moor" (A. J. P. Gore, Ed.), pp. 181–199. Elsevier, Amsterdam.

Backéus, I. (1987). Myten om tuvan som blev en hölja. *Fauna och Flora* **82**, 114–122.

Baker, J. H. (1972). The rate of production and decomposition of *Chorisodontium aciphyllum* (Hook. f. and Wils.) Broth. *British Antarctic Survey Bulletin* **27**, 123–129.

Barber, K. E. (1981). "Peat Stratigraphy and Climatic Change. A Palaeoecological Test of the Theory of Cyclic Peat Bog Regeneration." Balkema, Rotterdam.

Benda, I. (1957). Mikrobiologische untersuchungen über das auftreten von Schwefelwasserstoff in den anaeroben zonen des Hochmoores. *Archiv für Mikrobiologie* **27**, 337–374.

Boatman, D. J. (1983). The Silver Flowe National Nature Reserve, Galloway, Scotland. *Journal of Biogeography* **10**, 163–274.

Boatman, D. J., and Tomlinson, R. W. (1973). The Silver Flowe. I. Some structural and hydrological features of Brishie Bog and their bearing on pool formation. *Journal of Ecology* **61**, 653–666.

Bragg, O. M. (1982). "The acrotelm of Dun Moss—plants, water and their relationships." Unpublished Ph.D. thesis, University of Dundee.

Burgeff, H. (1961). "Mikrobiologie des Hochmoores." Fischer, Stuttgart.

Cajander, A. K. (1913). Studien über die Moore Finnlands. *Acta Forestalia Fennica* **2**(3), 1–208.

Casparie, W. A. (1969). Built-und Schlenkenbildung in Hochmoortorf. *Vegetatio* **19**, 146–180.

Chapman, S. B. (1964). The ecology of Coom Rigg Moss, Northumberland. I. Stratigraphy and present vegetation. *Journal of Ecology* **52**, 299–313.

Childs, E. C. (1969). "An Introduction to the Physical Basis of Soil Water Phenomena." Wiley, London.

Claricoates, J. (1990). "Gas production during peat decay." Unpublished Ph.D. thesis, University of London.

This chapter was written in spring 1988. Since that time there has been an efflorescence of work on methane evolution from peatlands, and our understanding of the interrelations of hydrology, size, and shape has improved.

Clark, R. L. (1986). The fire history of Rotten Swamp A.C.T. Australian Commonwealth Territory Parks and Conservation Service Report, pp. 1–21.

Clements, F. E. (1916). "Plant Succession. An Analysis of the Development of Vegetation." Carnegie Institution of Washington.

Clymo, R. S. (1963). Ion exchange in *Sphagnum* and its relation to bog ecology. *Annals of Botany (London)* NS **27**, 309–324.

——. (1965). Breakdown of *Sphagnum* in two bogs. *Journal of Ecology* **53**, 747–758.

——. (1983). Peat. *In* "Ecosystems of the World Vol. 4A. Mires: Swamp, Bog, Fen and Moor" (A. J. P. Gore, Ed.), pp. 159–224. Elsevier, Amsterdam.

——. (1984a). The limits to peat bog growth. *Philosophical Transactions of the Royal Society, London* B **303**, 605–654.

——. (1984b). *Sphagnum*-dominated peat bog: A naturally acid ecosystem. *Philosophical Transactions of the Royal Society, London* B **305**, 487–499.

——. (1988). A high-resolution sampler of surface peat. *Functional Ecology* **2**, 425–431.

Clymo, R. S., and Reddaway, E. J. F. (1971). Productivity of *Sphagnum* (bog-moss) and peat accumulation. *Hidrobiologia* **12**, 181–192. Also, without arbitrary cuts, as: "Aspects of the Ecology of the Northern Pennines." Moor House Occasional Paper No. 3. Nature Conservancy Council, Peterborough.

Collins, V. G., D'Sylva, B. T., and Latter, P. M. (1978). Microbial populations in peat. *In* "Production Ecology of British Moors and Montane Grasslands" (O. W. Heal and D. F. Perkins with W. M. Brown, Eds.), pp. 94–112. Springer-Verlag, Berlin.

Conway, V. M. (1948). Von Post's work on climatic rhythms. *New Phytologist* **47**, 220–237.

——. (1949). The bogs of central Minnesota. *Ecological Monographs* **19**, 173–206.

Coulson, J. C., and Butterfield, J. E. (1978). An investigation of the biotic factors determining the rates of plant decomposition on blanket bog. *Journal of Ecology* **66**, 631–650.

Crampton, C. B. (1911). The vegetation of Caithness considered in relation to the geology. Committee for the Survey and Study of British Vegetation, pp. 1–132.

Damman, A. W. H., and French, T. W. (1987). The ecology of the peat bogs of the glaciated northeastern United States: A community profile. U.S. Fish and Wildlife Service Biological Report 85 (7.16), pp. 1–100.

Dupont, L. M. (1985). "Temperature and rainfall variation in a raised bog ecosystem." Unpublished Ph.D. thesis, University of Amsterdam.

Eggelsmann, R. (1976). Moorhydrologie. *In* "Moor- und Torfkunde." (K-H Göttlich, Ed.), pp. 153–161. E. Schweitzerbart'sche, Stuttgart.

Eurola, S., Hicks, S., and Kaakinen, E. (1984). Key to Finnish mire types. *In* "European Mires" (P. D. Moore, Ed.), pp. 11–117. Academic Press, London.

Farrish, K. W., and Grigal, D. F. (1985). Mass loss in a forested bog: relation to hummock and hollow microrelief. *Canadian Journal of Soil Science* **65**, 375–378.

Fenton, J. H. C. (1980). The rate of peat accumulation in Antarctic moss banks. *Journal of Ecology* **68**, 211–228.

Field, E. M., and Goode, D. A. (1981). "Peatland Ecology in the British Isles: a Bibliography." Institute of Terrestrial Ecology and Nature Conservancy Council, London. Pp. 1–178.

Forrest, G. I., and Smith, R. A. H. (1975). The productivity of a range of blanket bog vegetation types in the northern Pennines. *Journal of Ecology* **63**, 173–202.

Foster, D. R., and Fritz, S. C. (1987). Mire development, pool formation and landscape processes on patterned fens in Dalarna, central Sweden. *Journal of Ecology* **75**, 409–437.

Foster, D. R., King, G. A., Glaser, P. H., and Wright, H. E., Jr. (1983). Origin of string patterns in northern peatlands. *Nature* (London) **306**, 256–258.

Foster, D. R., Wright, H. E., Jr., Thelaus, M., and King, G. A. (1988). Bog development and the dynamics of bog landforms in central Sweden and southeastern Labrador. *Journal of Ecology* **76**, 1164–1185.

Glaser, P. H. (1987). The ecology of patterned boreal peatlands of northern Minnesota: A community profile. U.S. Fish and Wildlife Service Biological Report 85 (7.14), pp. 1–98.

Glaser, P. H., and Janssens, J. A. (1986). Raised bogs in eastern North America: transitions in land forms and gross stratigraphy. *Canadian Journal of Botany* **64**, 395–415.

Glaser, P. H., Wheeler, G. A., Gorham, E., and Wright, H. E., Jr. (1981). The patterned mires of the Red Lake Peatland, northern Minnesota: vegetation, water chemistry and landforms. *Journal of Ecology* **69**, 575–599.

Gore, A. J. P., Ed. (1983). "Ecosystems of the World, Vols. 4A, 4B. Mires: Swamp, Bog, Fen and Moor." Elsevier, Amsterdam.

Gorham, E. (1953). Some early ideas concerning the nature, origin and development of peat lands. *Journal of Ecology* **41**, 257–274.

——. (1957). The development of peatlands. *Quarterly Review of Biology* **32**, 145–166.

Gorham, E., Santelmann, M. V., and McAllister, J. E. (1985). "A Peatland Bibliography, Chiefly with Reference to the Ecology, Hydrology and Geochemistry of *Sphagnum* Bogs." Department of Ecology and Behavioral Biology, University of Minnesota, Minneapolis. 152 pp.

Granlund, E. (1932). De Svenska högmossarnas geologi. *Sveriges Geologiska Undersökning Avhandligar och Uppsatser* **373**, 1–193.

Green, B. H., and Pearson, M. C. (1968). The ecology of Wybunbury Moss, Cheshire. 1. The present vegetation and some physical, chemical and historical factors controlling its nature and distribution. *Journal of Ecology* **56**, 245–267.

Harriss, R. C., Gorham, E., Sebacher, D. I., Bartlett, K. B., and Flebbe, P. A. (1985). Methane flux from northern peatlands. *Nature* (London) **315**, 652–654.

Heal, O. W., Flanagan, P. W., French, D. D., and Maclean, S. F., Jr. (1981). Decomposition and accumulation of organic matter in tundra. *In* "Tundra Ecosystems: A Comparative Analysis" (L. C. Bliss, O. W. Heal, and J. J. Moore, Eds.), pp. 587–663. Cambridge University Press, Cambridge.

Heal, O. W., and French, D. D. (1974). Decomposition of organic matter in tundra. *In* "Soil Organisms and Decomposition in Tundra" (A. J. Holding, O. W. Heal, S. F. Maclean, Jr., and P. W. Flanagan, Eds.), pp. 279–309. Tundra Biome Steering Committee, Stockholm.

Heal, O. W., Latter, P. M., and Howson, G. (1978). A study of the rates of decomposition of organic matter. *In* "Production Ecology of British Moors and Montane Grasslands" (O. W. Heal and D. F. Perkins with W. M. Brown, Eds.), pp. 136–159. Springer-Verlag, Berlin.

Heinselman, M. L. (1970). Landscape evolution, peatland types, and the environment in the Agassiz Peatlands Natural Area. *Ecological Monographs* **40**, 235–261.

Hulme, P. D. (1986). The origin and development of wet hollows and pools on Craigeazle Mire, south-west Scotland. *International Peat Journal* **1**, 15–28.

Ingram, H. A. P. (1978). Soil layers in mires: function and terminology. *Journal of Soil Science* **29**, 224–227.

——. (1982). Size and shape in raised mire ecosystems: a geophysical model. *Nature* (London) **297**, 300–303.

——. (1983). Hydrology. *In* "Ecosystems of the World Vol. 4A. Mires: Swamp, Bog, Fen and Moor" (A. J. P. Gore, Ed.), pp. 67–158. Elsevier, Amsterdam.

——. (1987). Ecohydrology of Scottish peatlands. *Transactions of the Royal Society of Edinburgh: Earth Sciences* **78**, 287–296.

Ivanov, K. E. (1953). "Gidrologiya Bolot." [Hydrology of Mires.] Gidrometeoizdat, Leningrad.

——. (1957). "Osnovy gidrologii bolot lesnoi zony i raschety vodnogo rezhima bolotnykh massivov." Gidrometeoizdat, Leningrad.

——. (1975). "Vodoobmen v Golotnykh Landschaftakh" [Water Movement in Mirelands]. Translated (1981) by A. Thomson and H.A.P. Ingram. Academic Press, London.

Janssens, J. A. (1983). A quantitative method for stratigraphic analysis of bryophytes in Holocene peat. *Journal of Ecology* **71**, 189–196.

Johnson, L. C., Damman, A. W., and Malmer, N. (1990). *Sphagnum* macrostructure as an indicator of decay and compaction on an ombrotrophic south Swedish peat bog. *Journal of Ecology*. **78**, 633–647.

Jones, H. E., and Gore, A. J. P. (1978). A simulation of production and decay in blanket bog. *In* "Production Ecology of British Moors and Montane Grasslands" (O. W. Heal and D. F. Perkins with W. M. Brown, Eds.), pp. 160–186. Springer-Verlag, Berlin.

King, W. (1685). On the bogs and loughs in Ireland. *Philosophical Transactions of the Royal Society, London* **15**, 948–960.

Kivinen, E., Haikurainen, L., and Pakarinen, P. (1979). Geographic distribution of peat resources and major peatland complex types in the world. *Annales Academiae Scientarum Fennicae*, Series A III Geologia-Geographica **132**, 1–28.

Kulczýnski, S. (1949). Peat Bogs of Polesie. Mémoires de l'Académie Polonaise des Sciences et des Lettres. Série B: 1–356.

Lindsay, R. A., Charman, D. J., Everingham, R., O'Reilly, R. M., and Rowell, T. A. (1988). "The Flow Country: The Peatlands of Caithness and Sutherland." Nature Conservancy Council, Peterborough.

Malmer, N. (1962). Studies on mire vegetation in the Archaean area of southwestern Götaland (south Sweden). 1. Vegetation and habitat conditions on the Åkhult Mire. *Opera Botanica* 7(1), 1–322.

Matthews, E., and Fung, I. (1987). Methane emisson from natural wetlands: Global distribution, area, and environmental characteristics of sources. *Global Biochemical Cycles* **1**, 61–86.

Middeldorp, A. A. (1984). "Functional paleoecology of raised bogs." Unpublished Ph.D. thesis, University of Amsterdam.

Moore, P. D., and Bellamy, D. J. (1974). "Peatlands." Elek Science, London.

Ohlson, M. (1987). Spatial variation in decomposition rate of *Carex rostrata* leaves on a Swedish mire. *Journal of Ecology* **75**, 1191–1197.

Osvald, H. (1923). "Die vegetation des hochmoores Komosse." Unpublished thesis, University of Uppsala.

Pearsall, W. H. (1956). Two blanket-bogs in Sutherland. *Journal of Ecology* **44**, 493–516.

Pearson, M. C. (1960). Muckle Moss, Northumberland. I. Historical. *Journal of Ecology* **48**, 647–666.

Rigg, G. B. (1940). The development of *Sphagnum* bogs in North America. *Botanical Reviews* **6**, 666–693.

——. (1951). The development of *Sphagnum* bogs in North America II. *Botanical Reviews* **17**, 109–131.

Rowell, T. K., and Turner, J. (1985). Litho-, humic- and pollen stratigraphy at Quick Moss, Northumberland. *Journal of Ecology* **73**, 11–25.

Siegel, D. I. (1983). Groundwater and the evolution of patterned mires, Glacial Lake Agassiz Peatlands, northern Minnesota. *Journal of Ecology* **71**, 913–921.

Siegel, D. I., and Glaser, P. H. (1987). Groundwater flow in a bog-fen complex, Lost River Peatland, northern Minnesota. *Journal of Ecology* **75**, 743–754.

Silvola, J. (1986). Carbon dioxide dynamics in mires reclaimed for forestry in eastern Finland. *Annales Botanicae Fennici* **23**, 59–67.

Sjörs, H. (1948). Myrvegetation i Bergslagen. *Acta Phytogeographica Suecica* **21**, 1–299.

——. (1961). Surface patterns in boreal peatland. *Endeavour* **20**, 217–224.

——. (1980). Peat on Earth: multiple use or conservation? *Ambio* **9**, 303–308.

Smith, R. I. Lewis, and Clymo, R. S. (1984). An extraordinary peat-forming community on the Falkland Islands. *Nature* **309**, 617–620.

Sturludottir, S. R., and Turner, J. (1985). The elm decline at Pawlaw mire: an anthropogenic interpretation. *New Phytologist* **99**, 323–329.

Svensson, B. H. (1980). Carbon dioxide and methane fluxes from the ombrotrophic parts of a subarctic mire. *In* "Ecology of a Subarctic Mire" (M. Sonesson, Ed.). *Ecological Bulletins* **30**, 235–250.

Svensson, B. H., and Rosswall, T. (1984). *In situ* methane production from acid peat in plant communities with different moisture regimes in a subarctic mire. *Oikos* **43**, 341–350.

Svensson, G. (1988). Fossil plant communities and regeneration patterns on a raised bog in South Sweden. *Journal of Ecology* **76**, 41–59.

Tallis, J. H. (1985). Mass movement and erosion of a southern Pennine blanket peat. *Journal of Ecology* **73**, 283–315.

——. (1987). Fire and flood at Holme Moss: erosion processes in an upland blanket mire. *Journal of Ecology* **75**, 1099–1129.

Tansley, A. G. (1939). "The British Islands and their Vegetation." Cambridge University Press, Cambridge.

Tolonen, K., Davis, R. B., and Widoff, L. W. (1983). Peat accumulation rates in selected Maine peat deposits. *Maine Geological Survey Bulletin* **33**, 1–97.

von Post, L., and Sernander, R. (1910). Pflanzen-physiognomische Studien auf Torfmooren in Närke. Livretguide des excursion en Suede du 11eme Congress Geologie Internationale **14**, 1–48.

Waksman, S. A., and Purvis, E. R. (1932). The microbiological population of peat. *Soil Science* **34**, 95–114.

Waksman, S. A., and Stevens, K. R. (1929). Contribution to the chemical composition of peat. 5. The role of microorganisms in peat formation and decomposition. *Soil Science* **27**, 315–340.

Walker, D., and Walker, P. M. (1961). Stratigraphic evidence of regeneration in some Irish bogs. *Journal of Ecology* **49**, 169–185.

Wallén, B. (1980). Structure and dynamics of *Calluna vulgaris* on sand dunes in south Sweden. *Oikos* **35**, 20–30.

Walter, H. (1977). The oligotrophic peatlands of western Siberia—the largest peinohelobiome in the world. *Vegetatio* **34**, 167–178.

Watt, A. S. (1947). Pattern and process in the plant community. *Journal of Ecology* **35**, 1–22.

Weber, C. A. (1908). Aufbau und vegetation der Moore Norddeutschlands. *Botanische Jahrbuch Suppl.* **90**, 19–34.

Wickman, F. E. (1951). The maximum height of raised bogs and a note on the motion of water in soligenous mires. *Geologiska Föreningens i Stockholm Förhandligar* **73**, 413–422.

Wright, H. E., Jr., and Glaser, P. (1983). Postglacial peatlands of the Lake Agassiz Plain, northern Minnesota. *In* "Glacial Lake Agassiz" (J. T. Teller and L. Clayton, Eds.), pp. 375–389. Geological Society of Canada, Special Paper 26, University of Toronto Press.

4
Paleoecology: Status and Prospect

James C. Ritchie

Paleoecology

The development of plant paleoecology as a rigorous discipline has always been impeded by a failure to make the phenomenological links to evolutionary biology and to community ecology. As a result, it has been largely excluded from the intellectual excitement and challenges of these fields. On the other hand, recent decades have seen fertile interactions with the earth sciences, propelled by technological advances in isotope chemistry and analysis, geochemistry, and large-scale computer simulations.

Boulding's description of an insecure science fits paleoecology exactly: "A field of knowledge is likely to be insecure if the available data only cover a small part of the total field and if the actual structures and relationships in it are extremely complex" (Boulding, 1980: 834). In part, of course, paleoecology is intrinsically impotent or "insecure," "simply because of the inadequacy of the historical record, which in the first place is a very small sample of the total field, and then is strongly biased in unknown directions by durability" (Boulding, 1980: 834).

This chapter reviews three topics of current interest in paleoecology where useful progress is being made, and speculates on those avenues of research that might hold the most promise in the immediate future. These topics are: (1) arid-land paleoecology, in which both investigations of modern desert ecology and refinements of packrat-midden macrofossil analysis are providing more secure bases for reconstructing past plant communities; (2) origin and evolution of floras, in which rates of evolution and modes of long-distance dispersal suggest useful reevaluations of old phytogeographical problems, particularly concerning origins of arctic floras; (3) the search for analogs, in which the application of analog analysis to large data banks of pollen, modern and fossil, is bringing spatial and temporal precision to searches for present-day plant communities that can be matched with fossil assemblages. The chapter will not touch on problematical themes that have been thoroughly reviewed in recent compendia, in particular the questions of vegetation-climate equilibrium (Webb, 1986; Birks, 1986; Woodward, 1987) and of the spatial and temporal scale of the pollen record as a tool in vegetation reconstruction (Birks, 1986; Delcourt and Delcourt, 1987, among others).

Arid Region Paleoecology

This chapter will examine Holocene environments, pollen signals, and vegetational responses to climatic change in arid North Africa, followed by reference to the southwestern United States as a region of potential significance in arid-land paleoecology.

Several lines of solid evidence have been assembled recently to confirm and elaborate the earlier suppositions, based on scattered, uncertain data, that areas of North Africa currently occupied by empty, hot deserts and influenced primarily by southwest monsoons, were cool and moist enough during the Holocene (10,000 to approx. 4,000 yr B.P.) to support diverse ecosystems and the presence of humans. The main classes of evidence are: former lake levels (Kutzbach and Street-Perrott, 1985); buried wadi systems (McCauley *et al.*, 1982; Pachur and Kropelin, 1987); buried lake sediments (Haynes *et al.*, 1979, 1989); invertebrate fossils (Haynes and Mead, 1987); faunal remains (Pachur and Kropelin, 1987); pollen (Ritchie and Haynes, 1987; Lézine, 1987); charcoal (Neumann, 1988); and archaeology (Wendorf and Schild, 1980). Climatic simulations by general-circulation models of the atmosphere based on Milankovitch mechanisms are consistent with the paleoecological evidence that a major pluvial episode occurred in North Africa during the Holocene, centered on the period 9,000 to 6,000 yr B.P. (Kutzbach and Guetter, 1986).

It appears that these few findings of early-Holocene pluvials fit broadly into the Croll-Milankovitch paradigms of orbitally forced, monsoon pulses (Prell and Kutzbach, 1987), but several major questions remain. First, the sedimentological evidence compiled so far shows that the initial lacustrine deposits were laid down as late as 9,000 yr B.P., though the postulated pluvial conditions driven by increased solar radiation and stronger monsoons, under conditions of a July perihelion and greater axial tilt, might be expected to show responses a few millennia earlier. The response lag might be a function of the rate of advance of a groundwater "front" as proposed by Haynes (1987).

Second, a major task remains to establish with precision the nature of both the pluvial climates and the vegetation cover. It is not surprising that interpretive procedures and paradigms developed in temperate regions are not completely effective at low latitudes. At least two factors of regional particularity must be accounted for: the spatiotemporal distribution of rainfall and the effects of biogeophysical feedbacks on aridity. Only a few cautionary intimations can be offered here. Let us examine possible modern analogs of early-Holocene, Saharan assemblages and see what can be learned in light of recent ecological investigations in the Negev desert (Evenari, 1986) and microclimatological analyses in east-central Sahara (Riou, 1975).

Spatiotemporal Distribution of Rainfall

In arid regions with rainfall of less than 100 mm, the rain is concentrated into a few months, varies greatly in space and time, and is significantly redistributed by runoff (Evenari, 1986). As a result, annual amounts of rainfall that are already quite small are sharply reduced, in terms of ecological availability of water, on rock plateaus and slopes, but greatly concentrated in wadis and lake basins. In arid regions, runoff is a complex process controlled by the intensity and duration of rainfall, the physical properties of the soil including temperature and water retention, and the nature of the surface materials of the landscape. Following Yair and Schachak (1982) and Schmida *et al.* (1986), Fig. 1 summarizes schematically the occurrence of different plant community–landscape complexes along gradients of habitat reliability (measured by the frequency of rainy years) and habitat favorability (measured as water availability). Although the scheme is largely theoretical, it illustrates that ecological responses to increases in rainfall (total and frequency of rainy years) will be nonlinear and landscape-dependent. Consequently a very small increase in annual rainfall (e.g., from 5 mm to 150 mm) might be accompanied by large, direct and indirect (groundwater recharge) responses in oases and wadis, and highly discontinuous responses in uplands. Thus, as the precipitation rises and water availability increases discontinuously, we might expect such ecological responses as: (1) increases of chenopods on poor, saline runoff (or infiltration) habitats (desert chenopods are

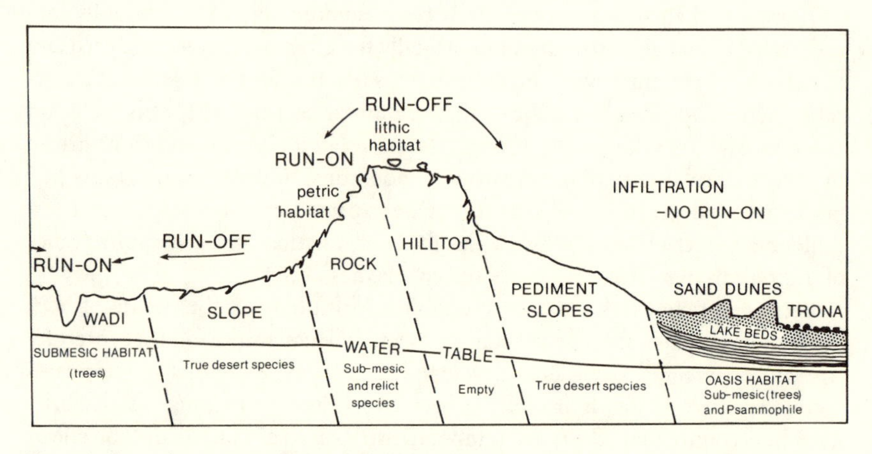

Fig. 1. A schematic summary of the main habitats of hot desert landscapes, their habitat characteristics, and major plant occurrences (based on Schmida *et al.*, 1986, Fig. 10).

perennial, summer-growth plants with C4 metabolism); (2) increases of tropical trees and shrubs in large wadi systems and oasis depressions; (3) increases of annuals and grasses on gentle slopes and in wadis; (4) increases of annuals and shrubs in small wadis.

A pollen record from the Selima oasis (Fig. 2), in the hyperarid core of the eastern Sahara (Fig. 3), shows some rough agreement with such a reconstruction. A detailed paleoenvironmental analysis is presented elsewhere (Haynes *et al.*, 1989), and the paleolimnological evidence suggests strongly that a relatively deep lake, up to 25 m deep at maximum with a chemically stratified water column, occupied the basin for several millennia. Pollen analysis of Holocene sediments from Saharan and other hyperarid regions is problematical because many of the dominant taxa of desert and savannah vegetation are low producers of pollen and, consequently, are underrepresented in pollen spectra. Nonetheless, the Selima record shows some correspondence with modern studies in the Negev, namely, that large proportions of pollen of Chenopodiaceae-Amaranthaceae and Gramineae are associated with small but consistently occurring percentages of such typical Saharo-Sahelian taxa as *Commiphora, Tribulus, Blepharis, Salvadora,* and *Maerua*.

The findings from the Selima project (Haynes *et al.*, 1989) and from other recent investigations in the Sahara (Gasse *et al.*, 1987; Lézine and Casanova, 1989) illustrate that pollen analysis alone is not effective in paleoecological reconstruction, but it can be more useful when combined with analysis of other biological and physicochemical proxy data.

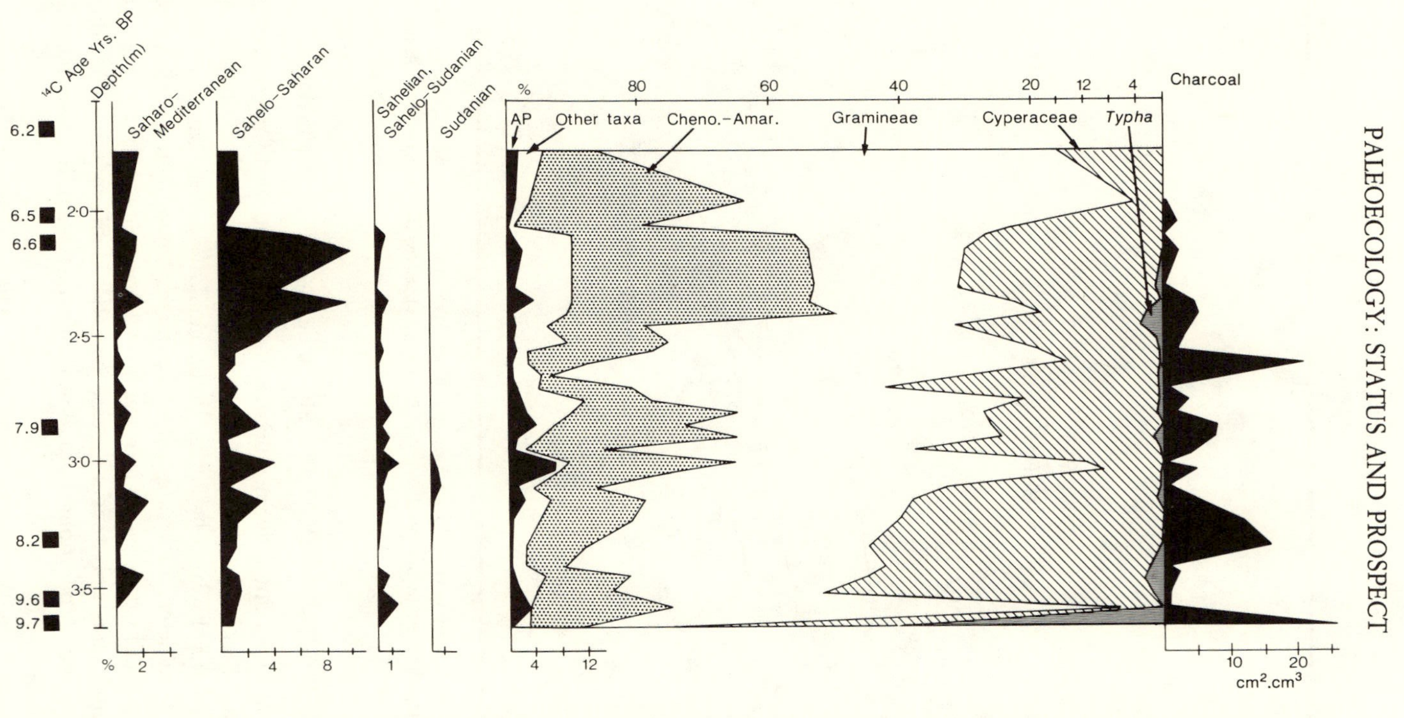

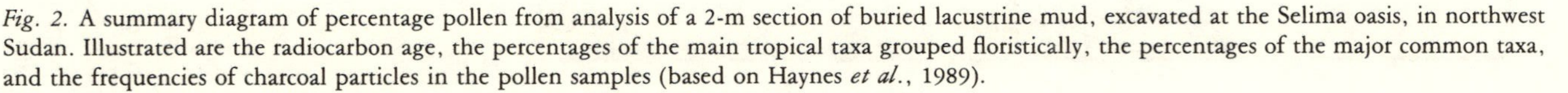

Fig. 2. A summary diagram of percentage pollen from analysis of a 2-m section of buried lacustrine mud, excavated at the Selima oasis, in northwest Sudan. Illustrated are the radiocarbon age, the percentages of the main tropical taxa grouped floristically, the percentages of the major common taxa, and the frequencies of charcoal particles in the pollen samples (based on Haynes *et al.*, 1989).

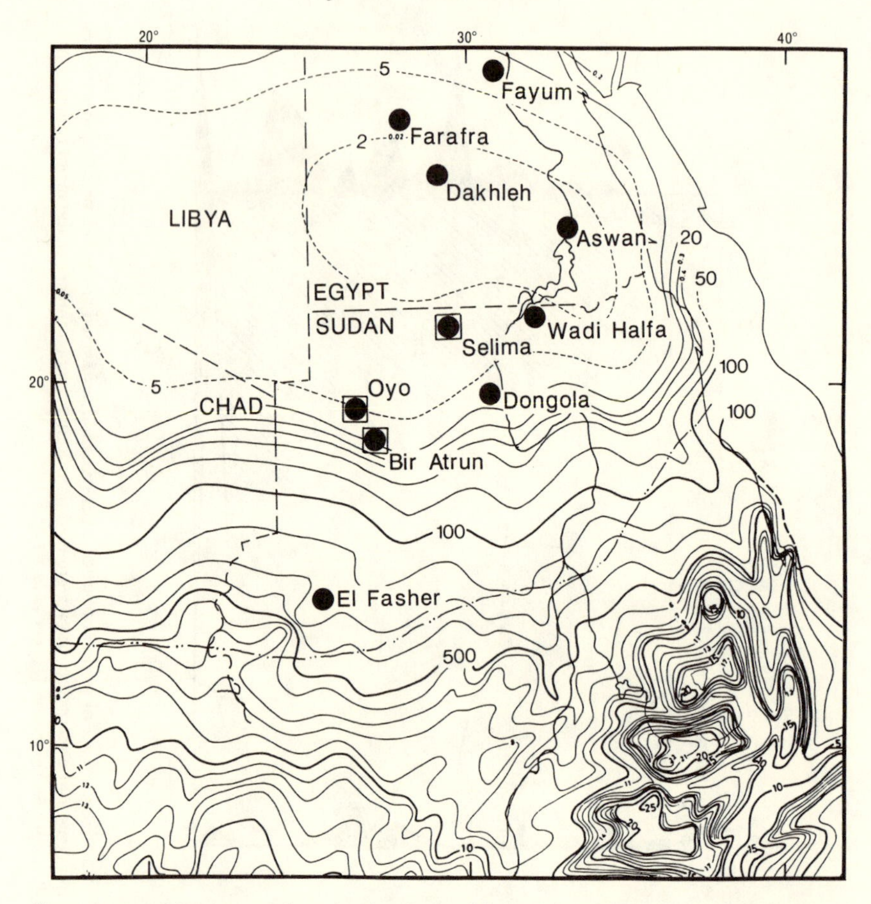

Fig. 3. A map of the eastern Sahara showing the localities mentioned in the text. The mean annual precipitation isohyets in mm are based on Leroux (1983). Isohyets above 500 are numbered in hundreds of mm.

Effects of Biogeophysical Feedbacks on Aridity

A second important difference between hot desert sites with no rainfall and those with between ca. 300 and 600 mm per annum is that, although the seasonal patterns of solar radiation are similar at all sites, even a slight increase in pluviosity in desert sites causes a decrease in mean monthly temperature, controlled in part by the effect of increases in water vapor in the atmosphere on net radiation. That is, only a slight increase in rainfall could produce marked differences in growing conditions, from those totally unfavorable to those able to sustain a diversity of plant life.

A detailed analysis of evapotranspiration along a transect of meteorological

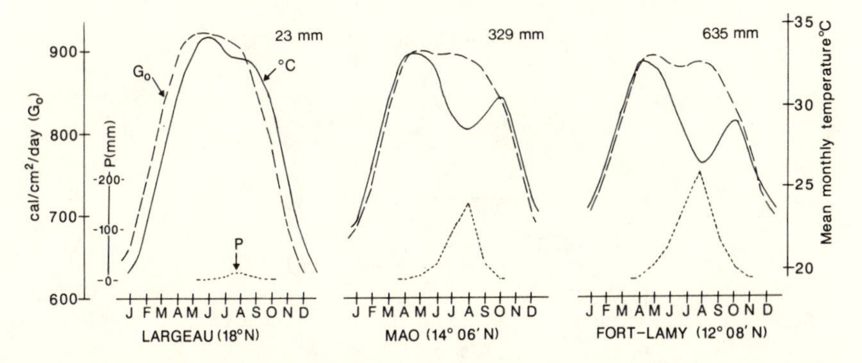

Fig. 4. The annual progression (1958 to 1968) of monthly ground-level values of net radiation (G_o), air temperature (°C) and precipitation (P) along a transect of sites from Largeau (absolute desert), Mao (desert steppe), and Fort-Lamy (sparse wooded savannah) in Chad. (Based on Riou, 1975, Fig. 81.) (With permission: Memoire 80, ORSTOM: 182.)

stations in Chad included one station in the southern Sahara with 23 mm annual precipitation and two in the northern Sahel with ca. 300 and ca. 600 mm, respectively (Fig. 4; based on Riou, 1975). A comparison among them might be a useful analog for the pluvial-hyperarid change in the Holocene. The effect of only a small difference in rainfall (< 150 mm) was to depress mean monthly temperature (means and maxima) by 4° C and approximately 8° C, respectively, as a result of the interaction of humidity and cloud with solar radiation (Fig. 4). Thus a very slight increase in rainfall might initiate significant changes in vegetation through a complex series of positive biogeophysical feedbacks as outlined in Fig. 5, based on Charney (1975). Increased rainfall promotes growth of vegetation cover, and the locally reduced values of albedo, combined with lower temperatures resulting from increased cloud and water vapor, might enhance local precipitation; however, it should be noted that Courel (1985) concludes from an analysis of albedo values for the Sahel for the years 1973 to 1979 that a positive correlation cannot be demonstrated between rainfall and albedo.

Along the Saharo-Sahel fringe in northwestern Sudan and adjacent regions of Chad, an arid region with mean annual precipitation < 50 mm, localized episodes of rainfall occur at decadal or shorter intervals. An instant vegetational response occurs, known locally as the gizu phenomenon. Many travellers and observers have recorded these transient plant communities and noted their importance in the economy of the nomadic pastoralists who live in this ecotone. Wilson (1978) notes that these plant communities, usually forming widely separated patches of 10 to 50 ha, consist of a few dozen species of annuals, rhizomatous perennials including several grasses, and a few shrubs. These species respond immediately to periods of a few days of rain by active vegetative growth, to form a patchy cover of scrub-grassland vegetation. As nomadic pastoral-

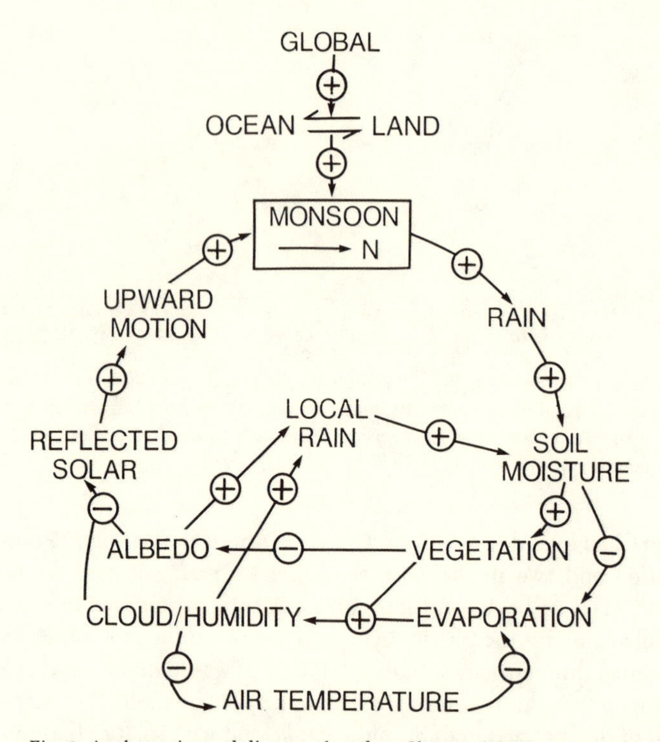

Fig. 5. A schematic word diagram, based on Charney (1975), to summarize possible interactions between increased annual precipitation, soil moisture, vegetation, evaporation, albedo, and local responses.

ism has become replaced by sedentary-settlement patterns during recent decades, the capacity of human populations to respond to the episodic appearances of patches of temporary grazing lands has diminished.

Charney (1975) has postulated that similar but negative biogeophysical feedbacks can enhance drought by the loss of vegetation and soil moisture, generating on the one hand, increased albedo and on the other, decreased cloud and humidity. The net result, in a weakly advective system such as the southeastern Sahara, is decreased upward motion "and the desert-monsoon interaction becomes self sustaining" (Charney 1975: 22). The importance of such an effect remains unmeasured, so the significance of changes in surfaces in reinforcing drought remains speculative (Nicholson, 1985).

The importance of these aspects of desert ecology for the paleoecologist are: (1) that only a slight increase in rainfall in an arid climate can produce significant vegetational responses; (2) that complex interactions between soil moisture, albedo, temperature, and locally increased rainfall can result; and (3) that small increments of rainfall can be redistributed on the landscape to produce enhanced effects in particular habitats.

The Arid Southwestern United States

The recent major increase in research in the semiarid regions of the southwestern United States (reviewed recently by Van Devender *et al.*, 1987) demonstrates the significant potential of the region for advancing paleoecology. Some of the central impediments encountered by pollen analysis in temperate regions disappear, but there are other, different problems. The most promising development is that the problems of temporal and spatial scale and of the taxonomic and floristic insensitivity of the proxy database are partly overcome by the analysis of packrat-midden macrofossils. The combination of such analyses with pollen and tree-ring investigations provides a formidable array of methods with which to confront the primarily biological problems of paleoecology, though difficulties persist in reconciling interpretations based on pollen data with those from macrofossil (midden) results. These conflicting interpretations are well illustrated by a recent exchange in print (Davis and Anderson, 1987; Van Devender *et al.*, 1987). The following brief summary of the potential of packrat-midden analysis is based on Van Devender *et al.* (1987) and Betancourt (1986).

It has been shown that packrat-midden assemblages are derived from local sources at the scale of plant communities. Packrats forage over areas within a 40- to 50-m radius of their midden sites. They retrieve >50% of the available species in the foraging range. Species can be identified in midden remains. As a result, detailed midden analyses can provide well-documented registrations of past plant communities. Recent investigations have shown that taxa that are indistinguishable morphologically can be differentiated by the DNA characteristics preserved in cellular extractions of the fossil material (Rogers and Bendich, 1985). Radiocarbon dating of discrete plant fragments by tandem-accelerator mass spectrometry has greatly improved the precision of packrat-midden chronologies.

Beyond the interests of community ecology and evolution, it is noteworthy that the sensitive registration in the southwest of global climatic change, effected by the El Niño Southern Oscillation, implies that the area has major importance for future investigations of paleoclimatic change (Betancourt, 1988). Equal promise holds at the other extremity of scale, for example, in the pinyon-juniper woodlands of the Southwest. Combined midden analysis, tree-ring studies, and age structure analyses make it possible for Betancourt to state that "not only can population growth be measured, but the pathways and rates of expansion, as well as the effects of competition, can also be observed directly" (Betancourt, 1986: 135).

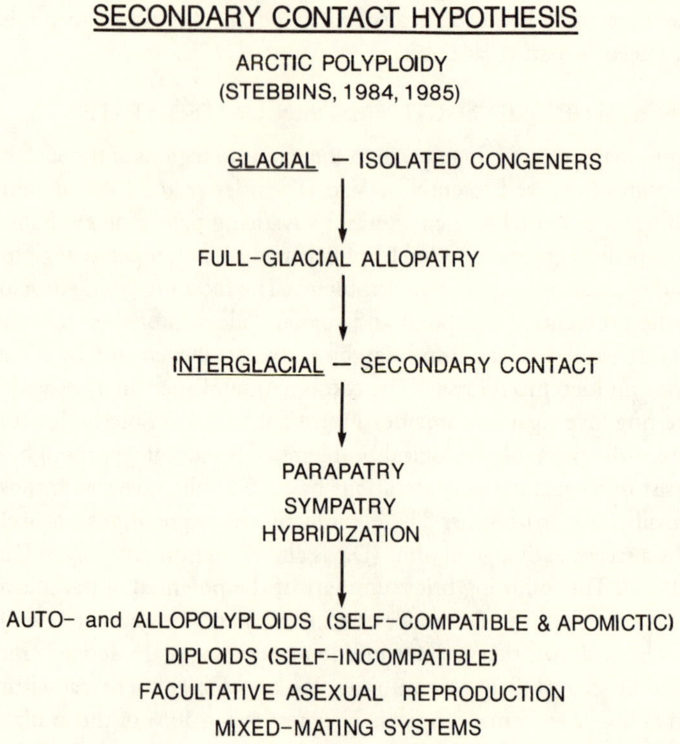

SECONDARY CONTACT HYPOTHESIS

ARCTIC POLYPLOIDY
(STEBBINS, 1984, 1985)

GLACIAL — ISOLATED CONGENERS

↓

FULL-GLACIAL ALLOPATRY

↓

INTERGLACIAL — SECONDARY CONTACT

↓

PARAPATRY
SYMPATRY
HYBRIDIZATION

↓

AUTO- and ALLOPOLYPLOIDS (SELF-COMPATIBLE & APOMICTIC)
DIPLOIDS (SELF-INCOMPATIBLE)
FACULTATIVE ASEXUAL REPRODUCTION
MIXED-MATING SYSTEMS

Fig. 6. A word diagram summarizing Stebbins's Secondary Contact Hypothesis. (Based on Stebbins, 1984, 1985, and Murray, 1987.)

Origin and Evolution of Floras

An old and recurring topic in phytogeography centers on the questions of origin of particular floras. Stebbins (1984, 1985) proposed that one effect of repeated Pleistocene glacial-interglacial cycles was to cause repeated disruptions and secondary contacts of plant populations. He has applied the idea in particular as a possible explanation of the patterns of genetic races and species in the arctic flora (Fig. 6). He suggests that the disruption of populations occurred during glacial periods. For example, at 18,000 yr B.P. the fossil record shows that such taxa as *Dryas, Oxytropis, Oxyria* (sorrel), *Salix* (willow), and *Saxifraga* (saxafrage) were disrupted into isolated populations around the Laurentide ice sheet. A detailed pollen and macrofossil analysis of the Wolf Creek site in Minnesota by Birks (1976) provides reliable evidence that several taxa common today in arctic regions were present south of the main Laurentide ice sheet. Similar assemblages have been recorded for the same time period in North Yukon and

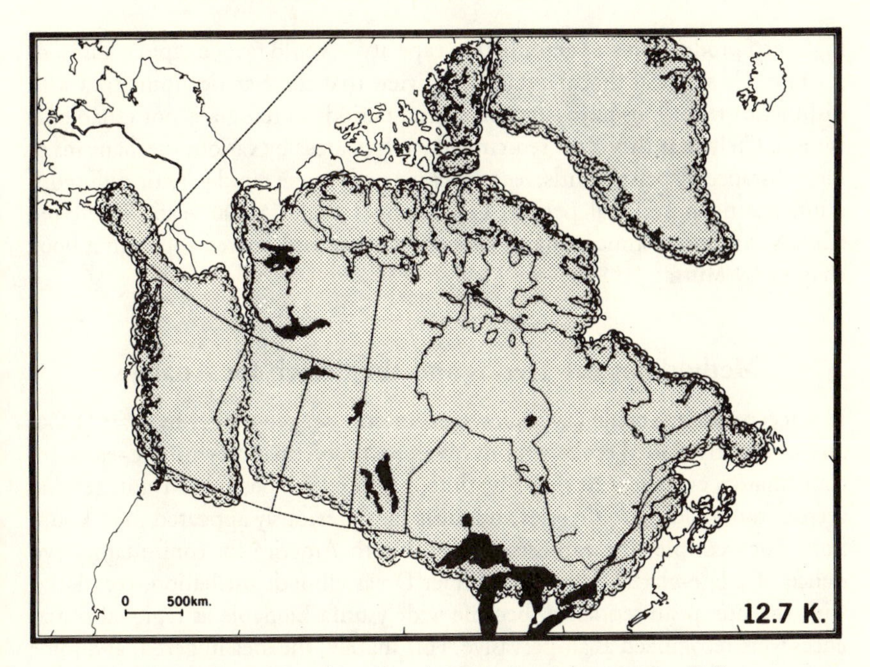

Fig. 7. The approximate position of ice sheets in North America at 12,700 yr B.P.

Alaska (reviewed recently by Lamb and Edwards, 1988). As the continental ice sheets decreased in area (Fig. 7), at least to the southeast, the pollen and macrofossil record documents a spread of these arctic taxa northward into freshly exposed landscapes (reviewed by Ritchie, 1987). Secondary contacts occurred in the late-glacial and Holocene as the ice sheets waned, and they promoted hybridization, often accompanied by "either polyploidy or introgression at diploid level" (Stebbins, 1984: 3). He concludes that "these new races and species, that now form the bulk of the arctic-alpine flora, originated during the entire period of a million years or more during which glaciers advanced and retreated, but some of them probably date from the beginning of the final recession, about 10,000 to 14,000 years ago" (Stebbins, 1984: 3).

An interesting, independent extension of these propositions has been suggested recently by Nordal (1987). He has reexamined an old phytogeographical problem—the arctic-alpine element in the vascular flora of Scandinavia—and concludes from an analysis of the genetic and reproductive structure of several taxa that "none of the . . . endemic species is necessarily very old. They may well be of no more than postglacial age (i.e. not older than about 15,000 years)" (Nordal, 1987: 378). He examines in particular the collective species *Papaver radicatum* (arctic poppy) and points out that its reproductive characteristics—

high seed production and germination capacity—would favor a rapid evolutionary rate. He suggests that the traditional view that these arctic-alpine taxa with disjunct distributions have survived glacial periods in refugia is not tenable. It is more likely that they have reoccupied ice-free areas by various mechanisms of long-distance dispersal (birds, ice) and have undergone rapid genetic differentiation during postglacial periods of no longer than 15,000 years. Interested readers can find a stimulating discussion of these and related topics in a book chapter by Murray (1987).

Methodological Trends and the Search for Analogs

In a recent essay, Loeble (1987) revisited themes set out earlier by Platt (1964) and has gently chided ecology for its obsessive preoccupation with the search for confirmatory evidence. At the same time, he noted the possible advantages that accrue from holding to theories and ideas that previously appeared to lack support. For example, the early searches in North America for confirmatory evidence of a late-glacial Allerød/Younger Dryas climatic oscillation, correlative with the European sequence, became widely unfashionable as regional differences were recognized as all-pervasive. Fortunately, the idea lingered, and some patient, observant investigators in eastern Canada finally assembled enough evidence to demonstrate the amphi-Atlantic coherency of the event (Mott *et al.*, 1986), and a recent, careful investigation has confirmed the sequence in portions of the midwestern Till Plains region (Shane, 1987). A serious impediment to establishing secure reconstructions of full-glacial and late-glacial vegetation, and thence of past environments, is that many pollen assemblages from these time spans lack equivalent or analogous modern spectra. Rapid improvements in computer technology have made relatively easy the task of comparing large numbers of modern pollen spectra with those from fossil samples. One product of the use of computer searches of large data banks to measure the similarity of fossil pollen assemblages with all possible modern spectra is the sharpening of the limits of paleoclimatic reconstruction (Overpeck *et al.*, 1985). Analog analyses provide mapped and down-core displays of the similarities between modern and fossil spectra, in spatial and temporal contexts. For example, such an application of this method to eight fossil sites in western North America—from which all fossil levels were compared with each of 300 modern spectra—produces a stark, spatial, and statistical statement of the familiar truth that late-glacial assemblages lack analogs (Anderson *et al.*, 1989).

On the other hand, such analyses can identify analogs of full-glacial pollen assemblages from northwestern Canada with sites in the modern midarctic (Fig. 8), and further confirm the mounting body of Alaskan evidence (Barnosky *et al.*, 1987) that bears out the conclusions of Cwynar (1982) that full-glacial Berin-

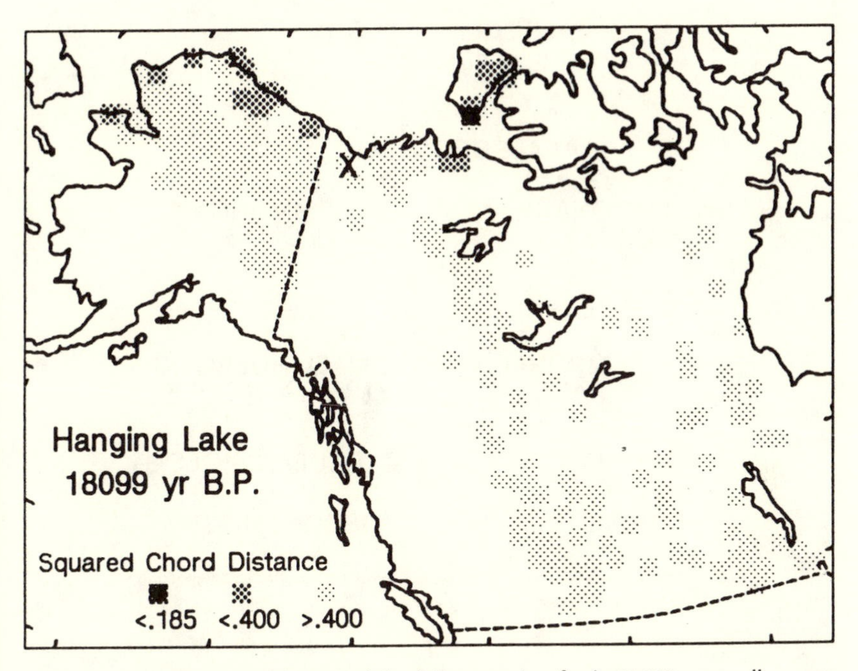

Fig. 8. A mapped summary of the squared chord-distance values for the 18,099 yr B.P pollen spectra at Hanging Lake (X), Yukon Territory, compared to each of a set of modern pollen spectra from Alaska and the western interior of Canada. Darker tones indicate closer similarities. Analogs for the 18,099 yr B.P. fossil sample are confined to northern Alaska and Banks Island.

gian landscapes were covered by sparse, herb-tundra communities similar to modern, midarctic Canadian ecosystems.

Although these new methods have greatly expanded the scope and speed of comparisons of modern with fossil spectra, the problems of elucidating the nature of vegetation change at the scale of plant communities are largely unsolved. Most pollen diagrams do not register clearly the changes related to local community dynamics because the source areas of the pollen signal are largely regions that include a wide range of vegetation types.

Conclusions

The main themes developed in this chapter are summarized in Fig. 9, but there is no attempt to provide a comprehensive review of the subject. It has been dealt with thoroughly in recent accounts by Birks and Birks (1980) and Birks (1986), and the points made here merely add and refine a few particular questions. In particular, if serious progress is to be made in linking paleoecology with commu-

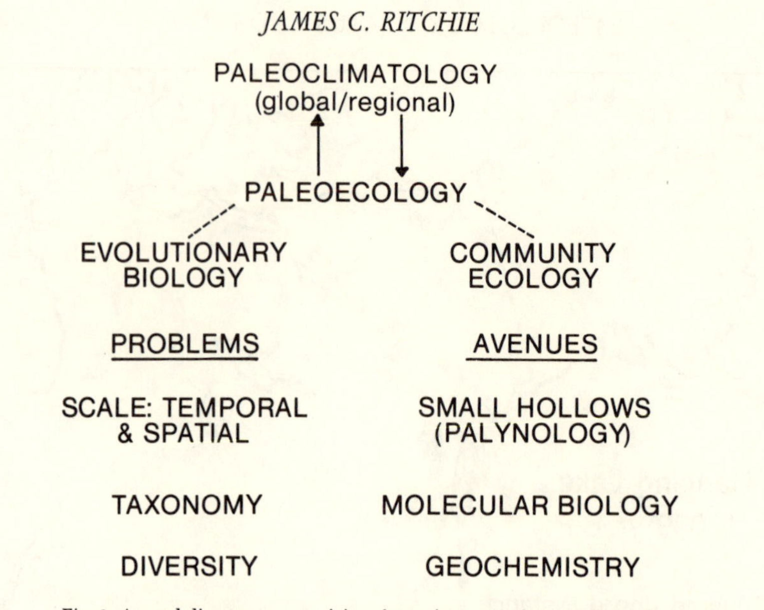

Fig. 9. A word diagram summarizing the main approaches to paleoecology discussed in this chapter.

nity ecology and evolution, it will be accomplished by close attention to questions of scale, taxonomic precision, and other lines of proxy data. Continued accumulation of regional-scale pollen diagrams will be useful in strengthening the linkage to paleoclimatology but ineffective in the direction of basic biological insights.

References

Anderson, P., Bartlein, P. J., Brubaker, L., Gajewski, K., and Ritchie, J. C. (1989). Late Quaternary climates of northwestern North America. 1. Modern analogues of late-Quaternary pollen spectra from the western interior of North America. *Journal of Biogeography* **16**: 573–596.

Barnosky, C. W., Anderson, P. M., and Bartlein, P. J. (1987). The northwestern U.S. during deglaciation; vegetational history and paleoclimatic implications. *In* "The Geology of North America, Vol. K-3, North America and Adjacent Oceans during the Last Deglaciation" (W. F. Ruddiman and H. E. Wright, Jr., Eds.), pp. 289–322. Geological Society of America, Boulder, Colo.

Betancourt, J. L. (1986). Palaeoecology of Pinyon-Juniper Woodlands; Summary. Proceedings of the Pinyon-Juniper Conference, January 13–16, Reno, Nev.

——. (1988). El Niño Southern Oscillation (ENSO) and Climate of the Southwestern U.S. Palaeoclimate Workshop, NSF & NOAA, February 15–17, Boston.

Birks, H. J. B. (1976). Late-Wisconsinan vegetational history at Wolf Creek, central Minnesota. *Ecological Monographs* **46**, 395–429.

——. (1986). Late-Quaternary biotic changes in terrestrial and lacustrine environments, with particular reference to north-west Europe. *In* "Handbook of Holocene Palaeoecology and Palaeohydrology" (B. E. Berglund, Ed.), pp. 3–65. Wiley, New York.

Birks, H. J. B., and Birks, H. H. (1980). "Quaternary Palaeoecology." Arnold, London.

Boulding, K. E. (1980). Science: Our common heritage. *Science* **207**, 831–836.

Charney, J. G. (1975). Dynamics of deserts and droughts in the Sahal. *Quaternary Journal of the Royal Meteorological Society* **101**, 193–202.

Courel, M-F. (1985). "Étude de l'évolution récente des milieux sahéliens a partier des esures fournies par les satellites." Thèse, l'Université Paris 1.

Cwynar, L. C. (1982). A late-Quaternary vegetation history from Hanging Lake, northern Yukon. *Ecological Monographs* **52**, 1–24.

Davis, O. K., and Anderson, R. S. (1987). Pollen in packrat (*Neotoma*) middens: Pollen transport and the relationship of pollen to vegetation. *Palynology* **11**, 185–198.

Delcourt, P. A., and Delcourt, H. R. (1987). "Long-term Forest Dynamics of the Temperate Zone: a Case Study of Late Quaternary Forests in Eastern North America." Springer-Verlag, New York.

Evenari, M. (1986). The desert environment. *In* "Ecosystems of the World Vol. 12B. Hot Deserts and Shrublands" (M. Evenari, I. Noy-Meir, and D. W. Goodall, Eds.), pp.1–22. Elsevier, Amsterdam.

Gasse, F., Fontes, J. C., Plaziat, J. C., Carbonel, P., Kaczmarska, I., de Deckker, P., Soulié-Marsche, I., Callot, Y., and Dupeuble, P. A. (1987). Biological remains, geochemistry, and stable isotopes for the reconstruction of environmental and hydrological changes in the Holocene lakes from North Sahara. *Palaeogeography, Palaeoclimatology, Palaeoecology* **60**, 1–46.

Haynes, C. V., Jr. (1987). Holocene migration rates of the Sudano-Sahelian wetting front, Arba'in Desert, eastern Sahara. *In* "Prehistory of Arid North Africa: Essays in Honor of Fred Wendorf" (A. E. Close, Ed.), pp. 60–84. Southern Methodist University Press, Dallas.

Haynes, C. V., Jr., Eyles, C. H., Pavlish, L. A., Ritchie, J. C., and Rybak, M. (1989). Holocene palaeoecology of the eastern Sahara: Selima Oasis. *Quaternary Science Reviews* **8**: 109–136.

Haynes, C. V., Jr., and Mead, A. R. (1987). Radiocarbon dating and palaeoclimatic significance of subfossil *Limicolaria* in northwestern Sudan. *Quaternary Research* **28**, 86–99.

Haynes, C. V., Jr., Mehringer, P. J., Jr., and Zaghloul, S. Z. (1979). Pluvial lakes of northwestern Sudan. *The Geographical Journal* **145**, 437–445.

Kutzbach, J. E., and Guetter, P. J. (1986). The influence of changing orbital parameters and surface boundary conditions on climate simulations for the past 18,000 years. *Journal of Atmospheric Sciences* **43**, 1726–1759.

Kutzbach, J. E., and Street-Perrott, F. A. (1985). Milankovitch forcing of fluctuations in the level of tropical lakes from 19 to 0 kyr. B.P. *Nature* **317**, 130–134.

Lamb, H. F., and Edwards, M. E. (1988). The arctic. *In* "Vegetation History" (B. Huntley and T. Webb III, Eds.), pp. 519–555. Kluwer Academic Publishers, Boston.

Leroux, M. (1983). "Le climat de l'Afrique tropicale." Chanpion, Paris.

Lézine, A-M. (1987). "Paléoenvironnements végétaux d'Afrique nord-tropicale depuis 1200 BP." Unpublished thesis, University of Aix-Marseille 2.

Lézine, A-M., and Casanova, J. (1989). Pollen and hydrological evidence for the interpretation of past climates in tropical West Africa during the Holocene. *Quaternary Science Reviews* **8**, 45–55.

Loeble, C. (1987). Hypothesis testing in ecology: Psychological aspects and the importance of theory maturation. *The Quarterly Review of Biology* **62**, 397–409.

McCauley, J. F., Schaber, G. G., Breed, C. S., Grolier, M. J., Haynes, C. V., Jr., Issawi, B., Elachi, C., and Blom, R. (1982). Subsurface valleys and geoarchaeology of the eastern sahara revealed by shuttle radar. *Science* **218**, 1004–1019.

Mott, R. J., Grant, D. R., Stea, R., and Occhiette, S. (1986). Late-glacial climatic oscillation in Atlantic Canada equivalent to the Allerød Younger Dryas event. *Nature* **323**, 247–350.

Murray, D. (1987). Breeding systems in the vascular flora of arctic North America. *In* "Differentiation Patterns in Higher Plants" (K. M. Urbanska, Ed.), pp. 239–262. Academic Press, New York.

Neumann, K. (1988). Die Bedeutung von Holzkohleuntersuchungen für die Vegetationsgeschichte der Sahara—das Biespiel Fachi-Niger. *Würzburger Geographische Arbeiten* **69**, 71–85.

Nicholson, S. E. (1985). Sub-Saharan rainfall 1981–1984. *Journal of Climate and Applied Meteorology* **24**, 1388–1391.

Nordal, I. (1987). *Tabula rasa* after all? Botanical evidence for ice-free refugia in Scandinavia reviewed. *Journal of Biogeography* **14**, 377–388.

Overpeck, J. T., Webb, T., III, and Prentice, I. C. (1985). Quantitative interpretation of fossil pollen spectra: dissimilarity coefficients and the method of modern analogs. *Quaternary Research* **23**, 87–108.

Pachur, H.-J., and Kropelin, S. (1987). Wadi Howar: Paleoclimatic evidence from an extinct river system in the southeastern Sahara. *Science* **237**, 298–300.

Platt, J. R. (1964). Strong inference. *Science* **146**, 347–353.

Prell, W. L., and Kutzbach, J. E. (1987). Monsoon variability over the past 150,000 years. *Journal of Geophysical Research* **92D**, 8411–8425.

Riou, C. (1975). La détermination pratique de l'évaporation. Application à l'Afrique Centrale. Memoire 80. ORSTOM, Paris. 236 pp.

Ritchie, J. C. (1987). "Postglacial Vegetation History of Canada." 175 pp. Cambridge University Press.

Ritchie, J. C., and Haynes, C. V., Jr. (1987). Holocene vegetation zonation in the eastern Sahara. *Nature* **330**, 645–647.

Rogers, S. O., and Bendich, A. J. (1985). Extraction of DNA from milligram amounts of fresh, herbarium and mummified plant tissues. *Plant Molecular Biology* **5**, 69–76.

Schmida, A., Evenari, M., and Noy-Mier, I. (1986). Hot desert ecosystems: An integrated view. *In* "Ecosystems of the World 12B. Hot Deserts and Shrublands" (M. Evenari, I. Noy-Mier, and D. W. Goodall, Eds.), pp. 379–387. Elsevier, Amsterdam.

Shane, L. C. K. (1987). Late-glacial vegetational and climatic history of the Allegheny Plateau and the Till Plains of Ohio and Indiana, U.S.A. *Boreas* **16**, 1–20.

Stebbins, G. L. (1984). Polyploidy and the distribution of the arctic-alpine flora: New evidence and a new approach. *Botanica Helvetica* **94**, 1–13.

——. (1985). Polyploidy, hybridation, and the invasion of new habitats. *Annals of the Missouri Botanical Garden* **72**, 824–832.

Van Devender, T. R., Thompson, R. S., and Betancourt, J. L. (1987). Vegetation history of the deserts of southwestern North America; the nature and timing of the late Wisconsin-Holocene transition. *In* "The Geology of North America, V. K-3. North America and Adjacent Oceans during the Last Deglaciation" (W. F. Ruddiman and H. E. Wright, Jr., Eds.), pp. 323–352. Geological Society of America, Boulder, Colo.

Webb, T., III (1986). Is the vegetation in equilibrium with climate? An interpretive problem for late-Quaternary pollen data. *Vegetatio* **67**, 75–91.

Wendorf, E., and Schild, R. (1980). "Prehistory of the Eastern Sahara." Academic Press, New York.

Wilson, R. T. (1978). The 'gizu': Winter grazing in the south Libyan Desert. *Journal of Arid Environments* **1**, 325–342.

Woodward, F. I. (1987). "Climate and Plant Distribution." Cambridge University Press, Cambridge.

Yair, A., and Schachak, M. (1982). A case study of energy, water and soil flow chains in an arid ecosystem. *Oecologia* **54**, 389–397.

5
Recent Paleolimnology and Diatom-Based Environmental Reconstruction

Richard W. Battarbee

Paleolimnology

In the last two decades paleolimnological studies have expanded rapidly, especially those concerned with the recent eutrophication and acidification of lakes. Paleolimnologists are making a unique contribution to the understanding of these problems because the sediment record enables past trends in water quality to be established on a large range of timescales (10^3–10^0 yrs) and because the record is often of a high quality, both in terms of its variety, richness, and temporal resolution.

Organic sediments occur in most lakes, usually accumulating in deep water (Fig. 1). The bulk of the material is derived from the catchment and the lake, but important trace components come from the atmosphere as well. The value and potential of the sediment record in studies of environmental change results

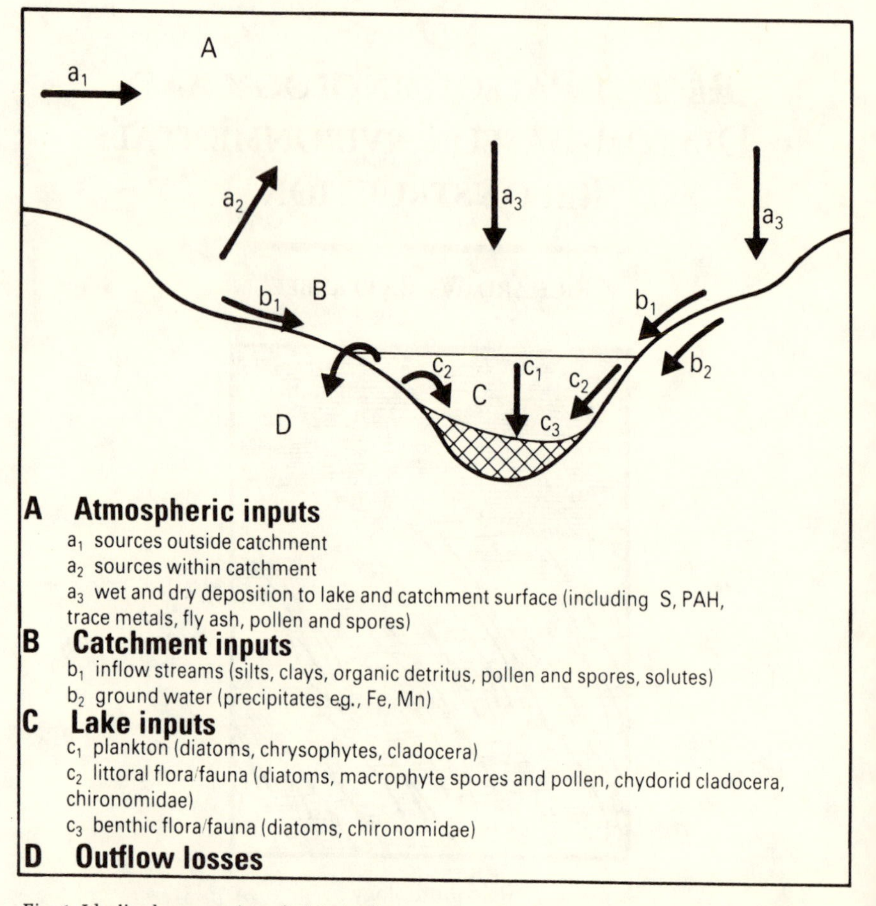

Fig. 1. Idealized cross section of a lake/sediment system showing sediment sources (from Battarbee *et al.*, 1988a). (With permission: ENSIS Ltd., London.)

from this diversity of source materials, which enables lake, catchment, and atmospheric histories to be unraveled, either separately or in combination.

The record is useful for a variety of reasons. Many scientists are interested in lake sediments but not necessarily in the lake and its history. For example, geomorphologists make use of lake sediments to reconstruct the history of catchment erosion (Dearing *et al.*, 1981); palynologists use them as a pollen trap (Bonny and Allen, 1984); and geophysicists use them to chart secular changes in the earth's magnetic field (Thompson, 1983). Even though these uses cannot be described as paleolimnology in the strict sense, all studies of this nature make contributions to paleolimnology directly through their considerations of sediment accumulation and chronology and, indirectly, through the generation of information that may help to explain limnological changes. Paleolimnology is concerned with an evaluation of the limnological history of lakes, rather than with the total information content of lake sediments.

This chapter explains the technical base of recent paleolimnology, indicates how diatom analysis in particular is and can be used for environmental reconstruction, and stresses that the study of the present and recent past is critical not just to enhance understanding of the impact of pollution but for the further development of paleolimnology in general.

Themes in Paleolimnology

ONTOGENY

Although some of the earliest paleolimnological research was devoted to pollution history (Nipkow, 1920), most early studies were more concerned with theories of lake development, especially in relation to trophic change (Deevey, 1942; Pennington, 1943; Hutchinson and Wollack, 1940), inspired in part by earlier attempts to classify lakes according to trophic status (Naumann, 1932; Thienemann, 1925) and to place them in an evolutionary series (Pearsall, 1921). These authors argued that the natural development of a lake was from oligotrophy to eutrophy and that the process should be called *eutrophication*. The term was inappropriate partly because many lakes, especially those in regions with slow-weathering bedrock, follow the opposite trend and become nutrient poor through time (Lundquist, 1927; Quennerstedt, 1955; Round, 1957; Digerfeldt, 1972) and partly because the term is confused with modern-day nutrient pollution (Rodhe, 1969). The neutral term *ontogeny* is, hence, preferred to describe the developmental history of lake systems (Whiteside, 1983).

Ontogeny is influenced by autogenic changes occurring as lake volume is reduced (Deevey, 1955), and by external changes in the character of soils and vegetation in the catchment under the influence of changing climate and human usage (Deevey *et al.*, 1979; Fritz, 1989). Important long-term changes in the ecosystem take place not only in productivity but also in acidity (Renberg and Hellberg, 1982; Whitehead *et al.*, 1986), color (Huttunen *et al.*, 1978), salinity (Bradbury *et al.*, 1981; Radle *et al.*, 1989), and other attributes. However, there are still too few studies of lake development to delimit the range of time trajectories experienced by lakes, and existing techniques for reconstructing some of the environmental variables are too primitive for trajectories to be defined with confidence. Progress in studies of lake ontogeny will be made by seeking appropriate modern analogs for different stages in lake history and by developing calibration sets of modern data suited to the reconstruction of crucial environmental variables, such as phosphorus, pH, color, and salinity.

Recent Human Impact

Few lakes in northern temperate regions have escaped human influences over the last 200 years or so. Although these influences contribute to the overall development of lakes, deflecting or reinforcing longer-term trends, it is convenient to treat these impacts separately. First, they can have an intense, direct, and rapid impact on lakes. Second, they can be identified and dated with greater certainty than changes occurring at early periods.

The most marked examples of recent pollution are eutrophication from nutrient enrichment (Bradbury, 1975; Battarbee, 1978a) and acidification caused by acid deposition (Battarbee, 1984; Charles and Norton, 1986; Davis, 1987). Soil, vegetation, and hydrological disturbances in lake catchments can also cause a range of limnological perturbations (Löffler, 1983; Liehu *et al.*, 1986).

The Rationale of Recent Paleolimnology

Recent sediments can be described conveniently as those that have accumulated within the last 200 years and are bounded at the upper limit by the mud-water interface. Working on this time-scale has special advantages:

1. The time period is most relevant for the study of contemporary environmental problems, the urgency of which stimulates the development of techniques and ideas for wider application.
2. Accurate sediment chronologies can usually be established, enabling data to be expressed as accumulation rates, if required.
3. The quality of fossil preservation can be assessed by comparison with the modern lake biota.
4. Modern-analog problems associated with the use of transfer functions are minimized.
5. Explanation of stratigraphic changes is maximized by the greater availability of well-documented historical records for lake and catchment change.
6. A quasi-experimental procedure can sometimes be adopted whereby the influence of known lake and catchment perturbations on the sediment record can be assessed—for example, the impact of a well-documented forest fire or a lake-liming event.
7. An understanding of the relationship between the recent sediment record and carefully observed or monitored processes in the lake and catchment forms the basis of a uniformitarian approach to the paleolimnology of all time periods.

Procedures: Sediment Sampling and Dating

Sampling

An emphasis on recent sediments has disadvantages as well as advantages for sediment coring and corer design. The record of the last 200 years or so is usually confined to the upper 1 m of sediment, so deeper, less penetrable sediments may sometimes be ignored. On the other hand, the uppermost sediment is often exceptionally fluid and must be sampled with great care and without causing disturbance to the mud-water interface.

The modified Livingstone corer (Wright, 1967) is probably the most versatile device for sampling deeper, consolidated organic sediments, but it cannot be used for the uppermost sediments because cores require horizontal extrusion. Instead, a simple corer based on the Livingstone design can be constructed. The drive rods are attached to a length of plastic tubing that is fitted with a piston. The corer is lowered to a suitable height above the sediment surface, the piston wire is secured, and the tube is driven past the piston (Wright, 1980). The core is then extruded vertically.

An alternative corer favored in the United Kingdom for this purpose is the mini-Mackereth corer (Mackereth, 1969), which is operated hydraulically and is particularly suited to deep water, windy conditions, and lakes that do not freeze over in the winter.

In situations where gas release causes disturbance of the surface sediment and when sampling at very close intervals is required, a freeze-corer is most appropriate. The first such corer was designed by Shapiro (1958). It has since been developed and popularized chiefly by Saarnisto (1979), Huttunen and Meriläinen (1978), and Renberg (1981a).

Dating

Perhaps the most important advantage of working with recent sediments is the potential precision and accuracy of dating techniques over this time period. These range from the globally applicable ^{210}Pb dating method (Goldberg, 1963; Krishnaswamy *et al.*, 1971) to the presence of lake-specific, datable stratigraphic markers (e.g., Digerfeldt *et al.*, 1975).

^{210}Pb

The ^{210}Pb content of sediment samples can be measured by wet-digestion techniques (Eakins and Morrison, 1978) or by direct gamma assay (Appleby *et al.*, 1986). The latter method is sometimes preferable since it is nondestructive and enables other isotopes to be measured at the same time. A typical gamma spectrum from a sample from Llyn Llagi is shown in Fig. 2.

^{210}Pb occurs naturally in lake sediments as one of the radioisotopes in the

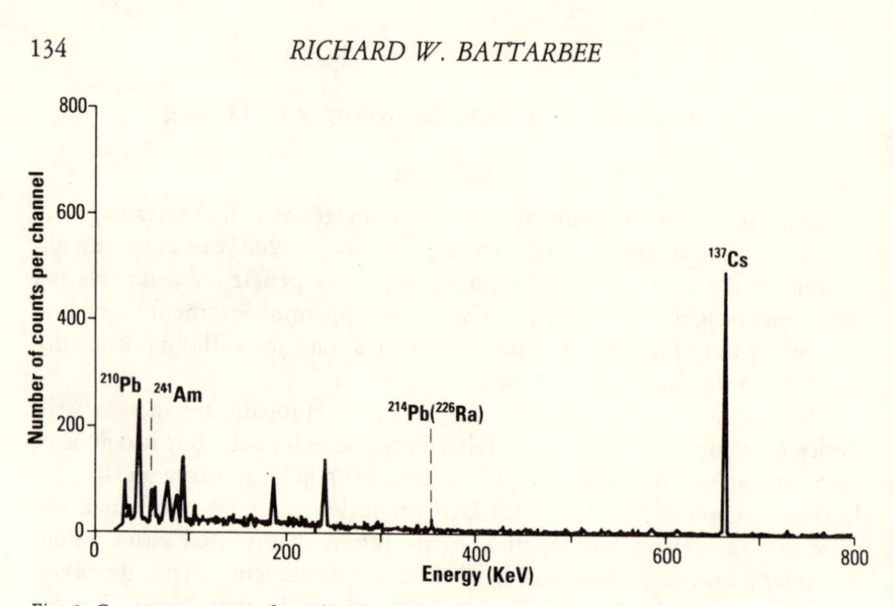

Fig. 2. Gamma spectrum of a sediment sample from Llyn Llagi, Wales (from Battarbee *et al.*, 1988a). (With permission: ENSIS Ltd., London.)

^{238}U decay series. It has a half-life of 22.26 years. The total ^{210}Pb activity in sediments has two components: "supported" ^{210}Pb derived from the in situ decay of the parent isotope ^{226}Ra; "unsupported" ^{210}Pb derived from the intermediate, gaseous isotope ^{222}Rn via an atmospheric pathway. In most samples the supported ^{210}Pb can be assumed to be in radioactive equilibrium with ^{226}Ra, so the unsupported activity at any level in a sediment core can be obtained by subtracting the ^{226}Ra activity from the total ^{210}Pb.

Figure 3 shows curves for the unsupported ^{210}Pb activity for two sites from which the age of each sample is calculated. In ideal situations where there are no catchment disturbances and the accumulation rate of dry mass of sediment is constant, the unsupported ^{210}Pb concentrations follows a log-linear decay curve with sediment depth. The concentration (C) in sediments of age t is then:

$$C = C(o)\, e^{-km/r}$$

where C(o) is the unsupported ^{210}Pb concentration at the sediment-water interface, k is the ^{210}Pb decay constant, m is the cumulative dry mass, and r is the sediment accumulation rate (Appleby and Oldfield, 1988).

In many situations the unsupported ^{210}Pb curve is nonlinear. For Llyn Llagi (Fig. 3a) the departure from linearity is not great, but for Loch Grannoch (Fig. 3b) a substantial soil inwash at a depth of 29 cm has caused a rapid increase in sediment accumulation and a consequent dilution of the ^{210}Pb concentration. In order to construct sediment chronologies from nonlinear profiles such as this, Appleby and Oldfield (1978) developed an alternative dating model on the as-

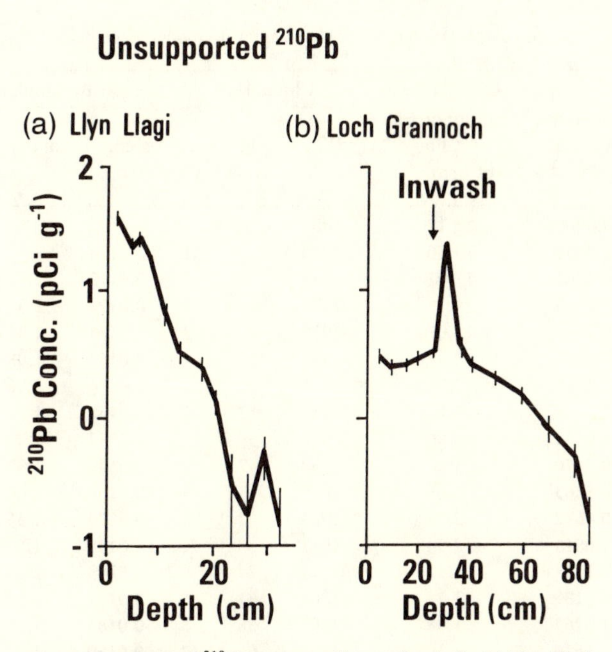

Fig. 3. Unsupported ^{210}Pb against sediment depth for (a) Llyn Llagi and (b) Loch Grannoch. Vertical lines are ± 1 standard deviation. pCi = picoCuries. (from Battarbee et al., 1988a). (With permission: ENSIS Ltd., London.)

sumption that the rate of supply of ^{210}Pb from the water column to the sediment was constant and independent of changes in the sediment-accumulation rate. This "constant-rate-of-supply" or "crs" approach is summarized by Appleby and Oldfield (1988) and Oldfield and Appleby (1984) as follows. If A denotes the unsupported ^{210}Pb inventory beneath sediments of depth x and A(o) denotes the ^{210}Pb inventory in the entire core, the age of sediment of depth x is given by:

$$A = A(o)\, e^{-kt}$$

where A and A(o) are calculated by direct numerical integration of the ^{210}Pb profile. The age of sediments of depth x is then given by

$$t = \frac{1}{k} \ln \frac{A(o)}{A}$$

Although the crs is often the one preferred, the basic assumption of the model is violated where there is evidence of significant sediment focusing, or where ^{210}Pb inputs from the catchment are relatively high. In these cases the simple model or a combination of models may be preferable. Appleby and Old-

Table 1. Chronology and accumulation rates for Llyn Llagi: CRS model (from Battarbee *et al.*, 1988a)

Depth cm	Cumulative dry mass g cm^{-2}	Cumulative unsupported ^{210}Pb pCi cm^{-2}	Chronology			Accumulation		
			Date a.c.e	Age yr	Error yr	g cm^{-3} yr^{-4}	mm yr^{-1}	Standard error %
0.0	0.0000	21.17	1985	0				
1.0	0.0548	19.30	1982	3	1	0.0174	2.62	5.2
2.0	0.1261	17.07	1978	7	2	0.0182	2.48	5.7
3.0	0.2045	14.95	1974	11	2	0.0191	2.37	6.3
4.0	0.2879	13.07	1970	15	2	0.0195	2.36	7.0
5.0	0.3710	9.48	1965	20	2	0.0167	1.98	6.6
6.0	0.4555	7.71	1959	26	2	0.0136	1.56	6.1
7.0	0.5438	6.22	1953	32	2	0.0132	1.39	6.4
8.0	0.6375	4.90	1946	39	2	0.0132	1.29	7.0
9.0	0.7470	3.86	1938	47	3	0.0146	1.37	8.7
10.0	0.8565	3.08	1930	55	3	0.0160	1.46	10.3
11.0	0.9672	2.54	1923	62	3	0.0174	1.55	11.9
12.0	1.0815	2.54	1917	68	4	0.0188	1.66	13.6
13.0	1.1958	2.10	1911	74	4	0.0203	1.77	15.3
14.0	1.3105	1.72	1904	81	5	0.0207	1.80	17.1
15.0	1.4267	1.38	1897	88	6	0.0183	1.58	19.6
16.0	1.5429	1.10	1890	95	7	0.0159	1.37	22.1
17.0	1.6591	0.89	1883	102	8	0.0135	1.16	24.6
18.0	1.7760	0.68	1875	110	10	0.0115	0.97	27.0
19.0	1.8952	0.47	1863	122	17	0.0093	0.76	31.2

field (1983) discuss procedures for selecting the most appropriate approach. In the case of Llyn Llagi the crs model performs well. Table 1 shows a full listing of dates, accumulation rates, and standard errors for this site, with data interpolated for each cm of core depth.

Fallout Isotopes from Nuclear Explosions

Lake sediments contain a record of radioisotope fallout from the testing of atomic weapons in the atmosphere. ^{137}Cs, with a half-life of about 30 years, has been used most frequently for dating. Ideally the shape of the ^{137}Cs activity versus sediment-depth profile should reflect the historically recorded fallout pattern (Fig. 4a). This begins in 1954, peaks in 1958/9, reaches a maximum in 1963, and declines rapidly to background values by 1968. Figure 4b shows a series of profiles for five lakes in the English Lake District (Pennington *et al.*, 1973) in which the general position of the 1963 maximum and the first main increase are shown. However, the post-1963 decrease in concentration seems delayed at some sites and, in the case of Blelham Tarn, there appears to be a downward tail of ^{137}Cs into older sediment. These departures from the ideal pattern can

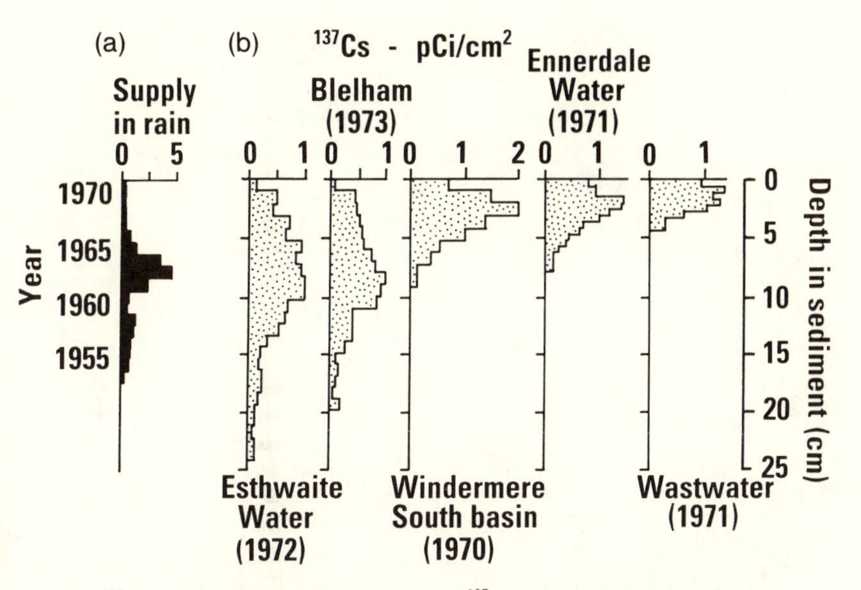

Fig. 4. (a) ^{137}Cs fallout in rain and (b) depth profiles of ^{137}Cs for five Cumbrian lakes (from Pennington *et al.*, 1973). (With permission: *Nature*; copyright 1973 Macmillan Magazines Ltd.)

be explained by sediment mixing (bioturbation and resuspension), delayed delivery of ^{137}Cs to the lake from the catchment, and diffusion of ^{137}Cs within the sediment column.

In acid lakes with highly organic sediments, ^{137}Cs seems quite mobile (Longmore *et al.*, 1983; Davis *et al.*, 1984). In these cases the stratigraphic pattern rarely reflects the fallout pattern. The peak can be at or very close to the sediment surface (later than 1963) and the beginning may be in sediment older than 100 years. In the case of Llyn Llagi (Fig. 5a) there is a significant ^{137}Cs concentration in sediments beyond the limit of detectable, unsupported ^{210}Pb (earlier than 1850).

In these situations $^{239,\,240}Pu$ (Jaakkola *et al.*, 1983) or ^{241}Am (Appleby and Oldfield, 1988) can provide useful alternatives, especially since both have very long half-lives in comparison with ^{137}Cs. ^{241}Am has a half-life of 432 years and is readily measured by gamma spectrometry. The ^{241}Am peak for Llyn Llagi (Fig. 5b) is not only in agreement with ^{210}Pb dating for this level but shows none of the mobility of ^{137}Cs.

Regional Stratigraphic Markers

Although the ^{210}Pb dating method can be used globally, it has a limited useful timespan. Because of the short half-life of ^{210}Pb, it is difficult to date periods before the middle of the 19th century with precision, yet by then many European lakes were strongly contaminated by trace metals from industrially derived

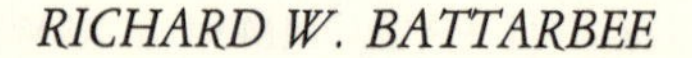

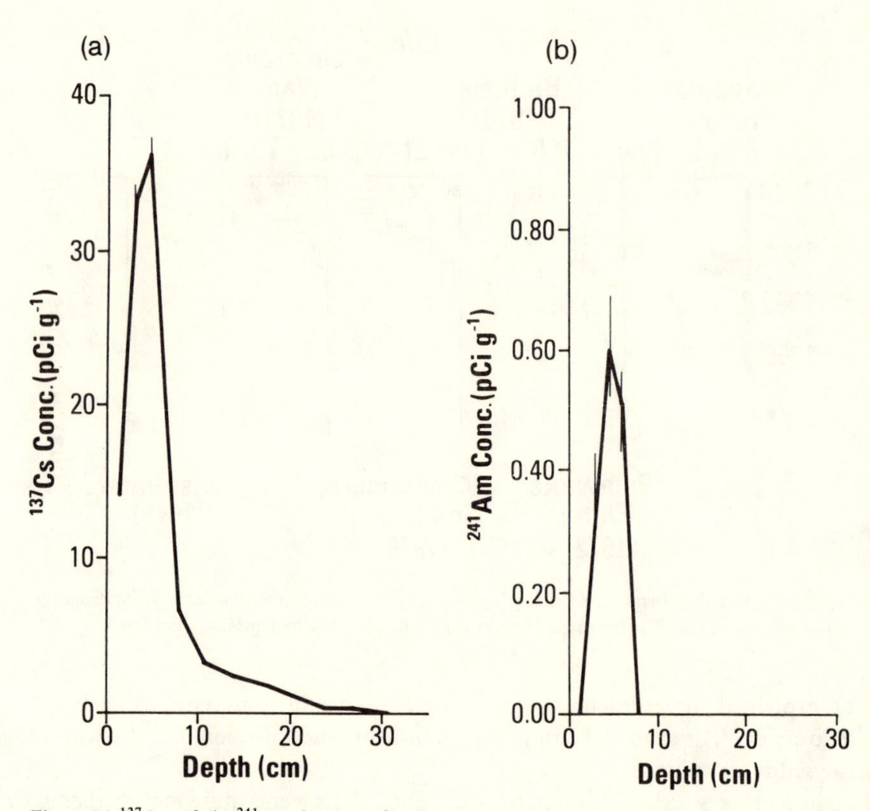

Fig. 5. (a) ^{137}Cs and (b) ^{241}Am depth profiles for Llyn Llagi (from Battarbee *et al.*, 1988a). (With permission: ENSIS Ltd., London.)

air pollution, and sensitive lakes were beginning to acidify (Battarbee *et al.*, 1988a). In the absence of a new isotopic technique for dating beyond the ^{210}Pb timescale, a regionally valid chronological system can be constructed from the stratigraphic record of pollen changes (e.g., the *Ambrosia* rise in the eastern United States) or of atmospheric contamination. In Britain the beginning of contamination of remote upland lakes by lead is recorded consistently in early-19th-century sediments (from extrapolation of the ^{210}Pb timescale). Within the ^{210}Pb timescale the appearance and rapid increase of spherical carbonaceous particles in sediments in the 1930s and 1940s are an excellent stratigraphic marker (Griffin and Goldberg, 1981, 1983; Renberg and Wik, 1984; Battarbee *et al.*, 1988a). Since most of these particles are derived from oil combustion, the horizon is approximately synchronous throughout western Europe and North America. Figure 6 shows Renberg and Wik's comparison of two profiles of

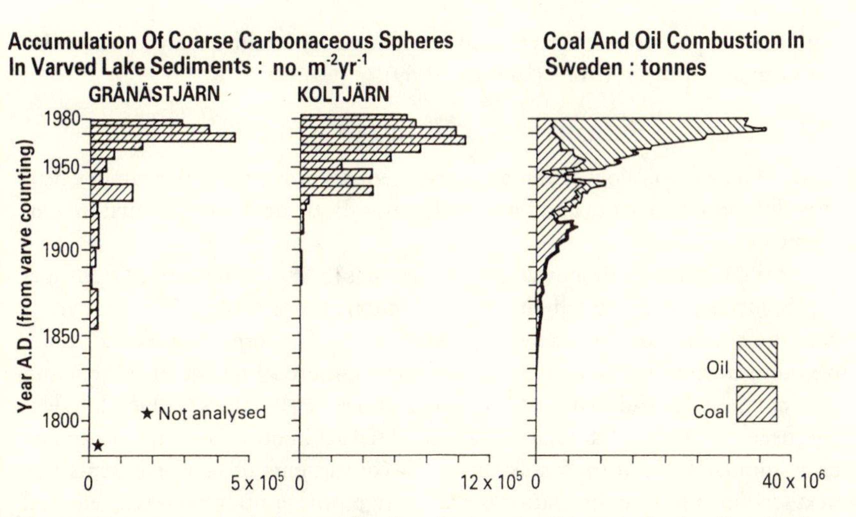

Fig. 6. Carbonaceous particle depth profiles from laminated sediments in Sweden (from Renberg and Wik, 1985). (With permission: *Ambio*.)

carbonaceous particles from lakes with varved sediments and the national statistics for oil and coal combustion in Sweden.

Site-Based Chronologies

With the use of freeze-corers it has become apparent that lakes with varved sediments are more common than previously thought (Saarnisto, 1979; Renberg, 1981b). Such sediments are typically found in meromictic lakes or relatively deep, dimictic lakes with anoxic hypolimnia. O'Sullivan (1983) has comprehensively reviewed the literature and puts forward a classification of varves according to their composition.

The most important consideration in using varves for dating purposes is to establish that they do indeed represent an annual cycle of deposition. This can be achieved in many ways including comparing results from this method with other dating systems, demonstrating seasonality from the microfossil succession within varves (Tippett, 1964; Simola, 1977; Battarbee, 1981a), and perhaps most simply, by repeat coring and showing the addition of a new varve each year (Renberg, 1986).

Other chronological information can be obtained from lithostratigraphic marker horizons that record lake and catchment events of known date. Nipkow (1920) shows inwash events that can be related to landslides of known age, and Digerfeldt *et al.* (1975) relate a series of clay lenses in the sediment of Järlasjön to the known development of an industrial estate on the shore of the lake. In some cases stratigraphic markers have been introduced deliberately into lakes for later use as dating horizons (Lawacz, 1969).

SEDIMENT-ACCUMULATION RATES

If an accurate chronology is available, sediment-accumulation rates can be calculated for individual cores and for a whole basin. Both are useful for environmental reconstruction.

For single cores, sediment-accumulation rates (SAR) can be derived from age-depth curves and expressed either in volumetric terms (mm yr^{-1} or cm yr^{-1}) or in dry-mass accumulation terms (g cm^{-2} yr^{-1}). The former is easier to visualize but is unsuitable for most recent sediments where a high water content and poor compaction lead to misleadingly high and variable values (Table 1). Calculations of sediment accumulation rate are essential in order to estimate the rate of accumulation of microfossils and other constituents of lake sediments (see next section) and as an indication of the stratigraphic sampling interval required to obtain a given temporal resolution.

Sediment chronologies can also be used to calculate whole-basin accumulation rates in situations where information is required on loss of reservoir capacity and where whole-lake budgets are being calculated (Evans and Rigler, 1980; Bloemendal *et al.*, 1979; Dearing, 1983). In simple situations, where it can be reasonably assumed that sediment has accumulated symmetrically (e.g., Lehmann, 1975) and conformably across a lake basin, it may be possible to predict sediment thickness, and thereby accumulation rate, at any point in the basin from water-depth data (Evans and Rigler, 1980; Engstrom and Swain, 1986). These conditions are likely to apply in small, relatively deep, seasonally stratified continental lakes with winter ice cover where sediment is focused mainly by gravitational forces to deeper water from all edges.

At the other extreme are exposed, isothermal lakes in oceanic environments with no winter ice cover where sediment accumulates in sheltered, often more marginal locations in the basin or is even lost from the basin. In these cases sediment is distributed asymmetrically (thickest sediment does not equate with maximum water depth), hiatuses (both single-core and whole-basin) may be common, and accumulation-rate patterns are sensitive to changes in the lake environment, especially shelter (Anderson *et al.*, 1986; Battarbee, unpublished).

Between these limits all gradations probably occur with the focus of sediment accumulation shifting in space and time (Battarbee, 1978b; Dearing, 1983). In these latter situations, whole-basin accumulation rates cannot be predicted from simple assumptions that relate accumulation rates to water depth. Instead the space-time pattern can be established from a stratigraphic comparison of a number of cores in which all cores are dated (e.g., Engstrom and Swain, 1986) or in which a chronology from a designated master-core is transferred to other cores by various stratigraphic correlation procedures (Dearing, 1986; Anderson, 1986a).

Information Content of Lake Sediments

Although lake sediments contain rich and varied fossil assemblages, there have been relatively few systematic studies of the extent to which these quantitatively and qualitatively represent the structure and productivity of lake communities from which they are derived. This question of representativity relates not only to the taphonomy of preserved material but also to the lack of representation of entire biotic groups. For example, the quality of preservation varies to such an extent that few copepods or green algae are preserved at all, whereas for diatoms and chydorid cladocera it is often possible to generate a more complete species list from samples of surface mud than from samples of live communities.

Continued research is needed (1) to evaluate the quality of the record of well-preserved and routinely used groups, such as diatoms, cladocera, and chironomids, and (2) to develop new direct or indirect techniques to fill information gaps. Diatoms, cladocera, and chironomids are the most frequently used paleolimnological fossil groups since they tend to be well preserved, diverse, ubiquitous, and identifiable to a low taxonomic level. In some cases intact diatom frustules, entire cladocera, and unbroken chironomid head capsules occur in sediments. Normally, however, material is disarticulated and broken, and identification requires skill and experience. A more severe problem, especially for diatomists, is dissolution, which is differential between species and between sites (Battarbee, 1986).

For recent sediments the quality of representation can be assessed by comparing fossil assemblages with modern or historical algal records. Several studies have shown that the relationship between diatom-plankton communities and their representations in the sediment record can be very close (Simola, 1977; Battarbee, 1979, 1981b; Haworth, 1980). Figure 7 shows such a comparison for Blelham Tarn (Haworth, 1980), where an almost-perfect match is demonstrated between the historical algal and the sediment records. On the other hand, in a similar comparison for Lough Neagh, Battarbee (1979) showed that the long, fragile diatoms *Synedra acus* and *Fragilaria crotonensis* were less common in the 1900–1910 sediment than would have been predicted from contemporary plankton collections (Dakin and Latarche, 1913).

It might be expected that periphytic diatom populations are less well represented in deep-water sediments. In a study of the diatom periphyton of the Round Loch of Glenhead, Jones and Flower (1986) showed good agreement between the proportions of the dominant taxa in the periphyton and in a surface-sediment sample but interesting disagreements in taxa with lower abundance

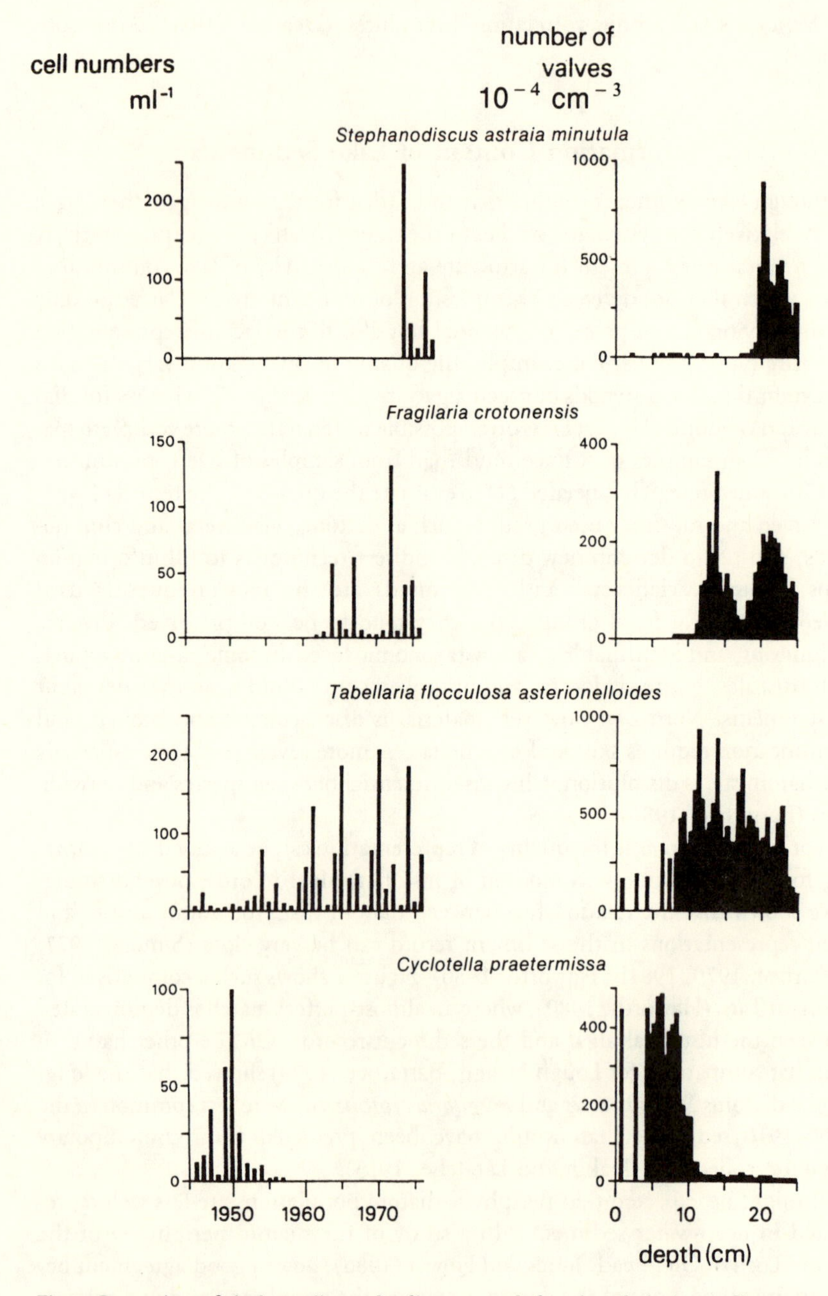

Fig. 7. Comparison of algal sampling and sediment records for selected planktonic diatoms from Blelham Tarn (redrawn from Haworth, 1980). (With permission.)

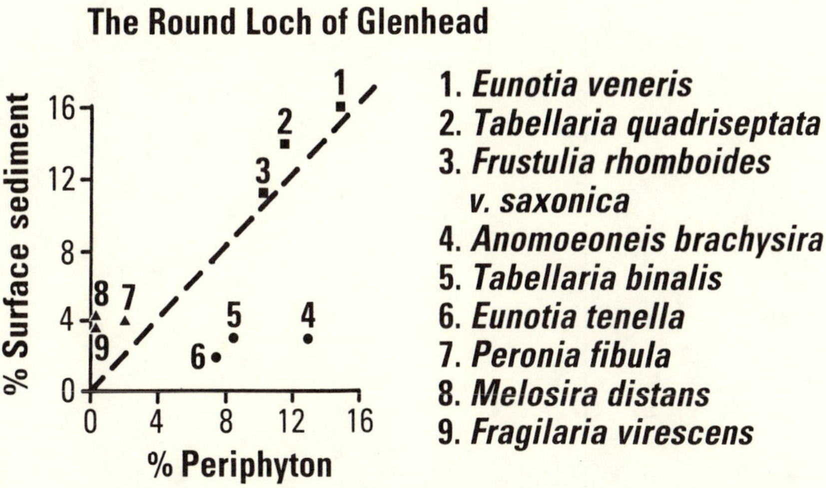

The Round Loch of Glenhead

% Surface sediment

% Periphyton

1. *Eunotia veneris*
2. *Tabellaria quadriseptata*
3. *Frustulia rhomboides v. saxonica*
4. *Anomoeoneis brachysira*
5. *Tabellaria binalis*
6. *Eunotia tenella*
7. *Peronia fibula*
8. *Melosira distans*
9. *Fragilaria virescens*

Fig. 8. Comparison of modern periphyton samples with the sediment assemblage for the Round Loch of Glenhead (data from Jones and Flower, 1986). (With permission.)

(Fig.8). In some cases the periphyton samples have higher proportions, in others (e.g., *Melosira distans* and *Fragilaria virescens*) the surface sediment contains 4% but neither taxon was collected in any frequency in live material. This is either because not all microhabitats in the lake were sampled or because these taxa were reworked from older sediments. Overall the agreement is quite close and indicates that periphytic diatoms in small lakes are effectively transported from marginal habitats to the point from which cores are normally taken. At sites where the assemblage is dominated by planktonic diatoms, a greater total count is often needed to derive accurate proportions of periphytic taxa, and individual counts are better expressed as a percentage of the sum of periphyton rather than the sum of all diatoms.

Many other groups of organisms are represented in lake sediments, though in most cases the record is incomplete or more difficult to interpret in some way. Chrysophyte cysts are well preserved and extremely abundant, but for most it is impossible to identify species. Of the 450 morphotypes now described from sediments, only 30 have been referred to species (Kristiansen, 1986). Nevertheless, progress is being made following the formation of the International Statospore Working Group, which has proposed guidelines for describing cysts (Cronberg and Sandgren, 1986), and occasionally a definite link is made between cyst and parent from observations of cyst liberation. In this way Cronberg (1973) showed that Nygaard's *Cysta teres* (Nygaard, 1956) belongs to *Mallomonas eoa*.

The use of chrysophyte scales, especially from the Mallomonadaceae, has

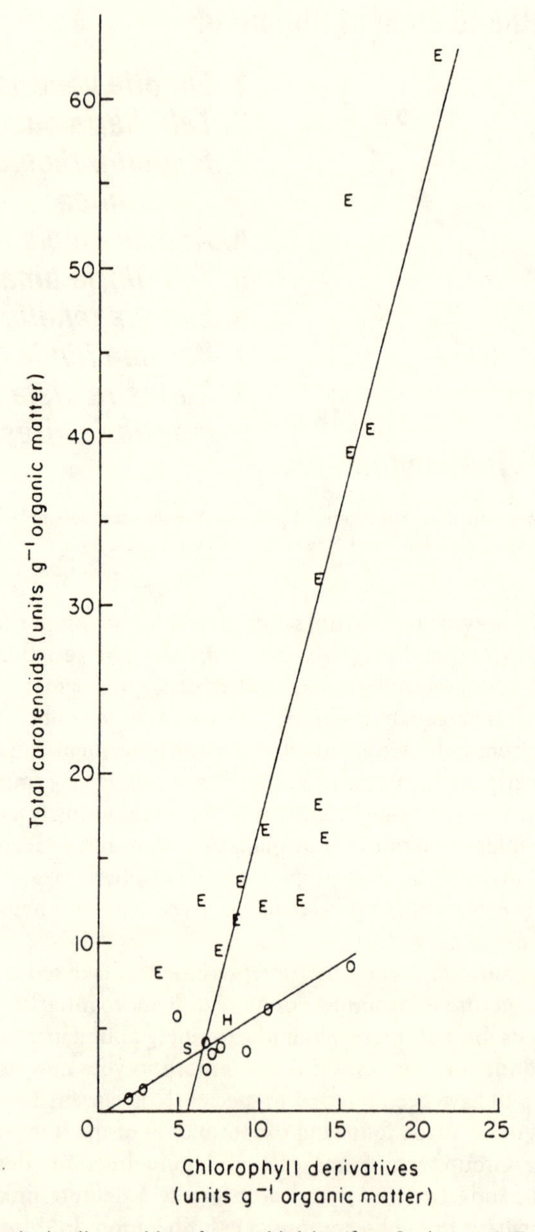

Fig. 9. Pigment ratios in oligotrophic and eutrophic lakes (from Swain, 1985). E = eutrophic lakes as described in original article; O = oligotrophic lakes of the Boundary Waters Canoe Area, Minnesota; S = Shagawa Lake, Minnesota; H = Harvey's Lake, Vermont. (With permission: Blackwell Scientific Publications Limited.)

more potential (Smol, 1980; Munch, 1980; Battarbee *et al.*, 1980). In most cases scales can be reliably identified, and they often occur in sufficient concentration and diversity to be useful for environmental reconstruction (Smol *et al.*, 1984; Hartmann and Steinberg, 1986).

Nonsiliceous algal remains also occur in lake sediments. Of these the *Pediastrum* (Chlorococcales) group is the best preserved (Cronberg, 1982), but in suitable conditions a range of green, blue-green, and dinoflagellate vegetative cells and resting spores also can be found (Livingstone, 1984; Cronberg, 1986). Pollen, spores, and seeds of aquatic plants are also common and have been extensively used in studies of water level and climatic change (Digerfeldt, 1986; Watts and Winter, 1966), but have been used only rarely for water-chemistry reconstruction, despite the pioneering study by Birks (1973), who compared modern macrofossil assemblages to water quality and climatic gradients in Minnesota.

Animal microfossil techniques also have great potential. Many of the animal groups found in lake sediments (Frey, 1964) have not been fully explored, though new progress is being made with, for example, rhizopoda (Douglas and Smol, 1987) and Bryozoa (Crisman *et al.*, 1986). Interest in ostracods has continued. Löffler (1986a) demonstrated their value in indicating the onset of meromixis in alpine lakes (Löffler, 1975, 1986b) Chivas *et al.* (1986) and Engstrom and Nelson (1991) showed recently that variations in the magnesium and strontium content of ostracod shells can be used to reconstruct salinity and temperature.

Primary producers in lake systems do not always leave morphological traces in lake sediments but may contribute to the fossil pigment record (Fogg and Belcher, 1961; Gorham and Sanger, 1972). Although chlorophyll-degradation products in sediments are derived from all primary producers, the carotenoid pigments oscilloxanthin and myxoxanthophyll are more specifically related to blue-green algal populations (Züllig, 1961, 1982; Griffiths, 1978; Engstrom *et al.*, 1985). However, interpretation of pigment concentrations and accumulation rates in terms of changing productivity and population structure can be ambiguous and requires a detailed knowledge of the factors influencing pigment preservation in sediments (Swain, 1985). The importance of this understanding is illustrated by Swain (1985), who compared the ratio of chlorophyll derivatives to total carotenoids in the surface sediments of a group of oligotrophic lakes with those in a group of eutrophic lakes. Figure 9 shows different slopes for the two groups, with a higher carotenoid-to-chlorophyll ratio in eutrophic lakes. From within-lake transect studies, Swain argues that this results from a faster degradation rate of total carotenoids in oligotrophic lakes rather than a greater production of carotenoids in eutrophic lakes.

At the present time only a few of these analytical approaches can be used in isolation for confident environmental reconstruction. Individually many are

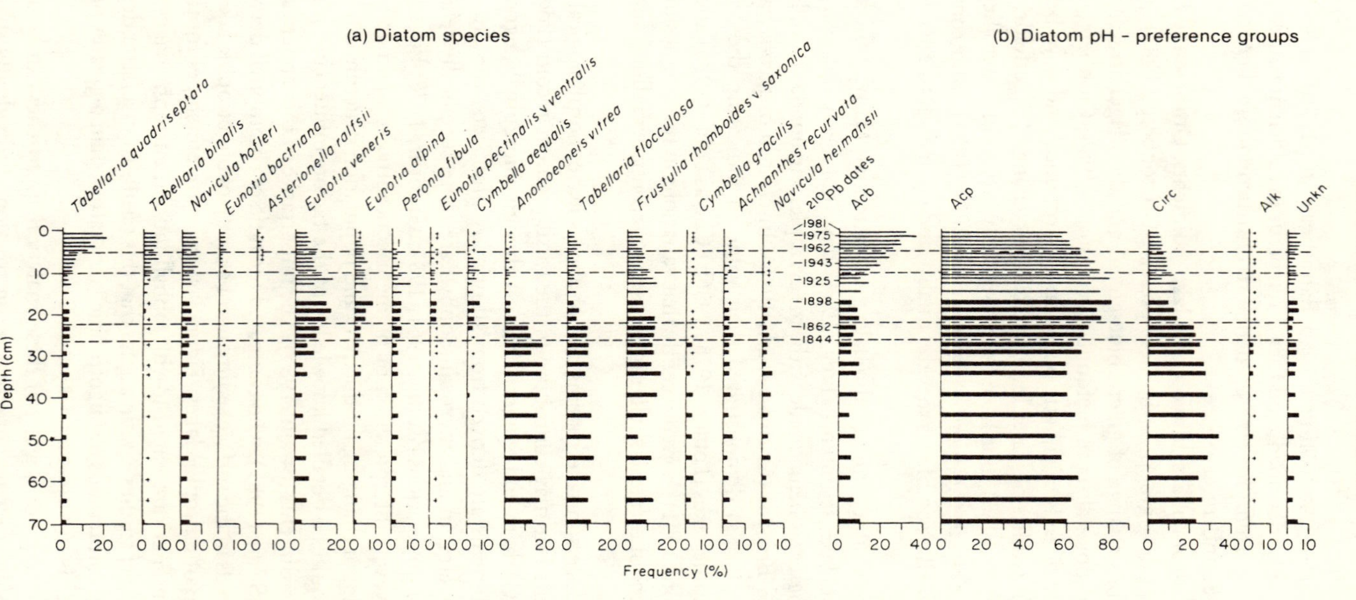

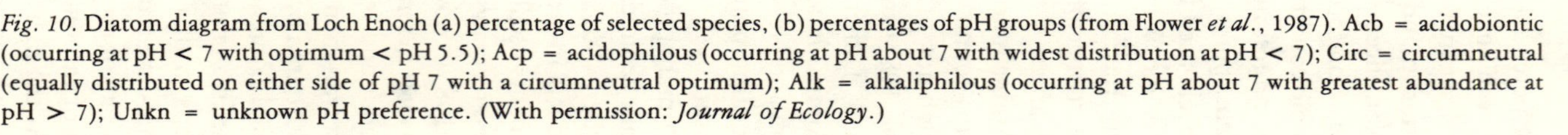
Fig. 10. Diatom diagram from Loch Enoch (a) percentage of selected species, (b) percentages of pH groups (from Flower *et al.*, 1987). Acb = acidobiontic (occurring at pH < 7 with optimum < pH 5.5); Acp = acidophilous (occurring at pH about 7 with widest distribution at pH < 7); Circ = circumneutral (equally distributed on either side of pH 7 with a circumneutral optimum); Alk = alkaliphilous (occurring at pH about 7 with greatest abundance at pH > 7); Unkn = unknown pH preference. (With permission: *Journal of Ecology*.)

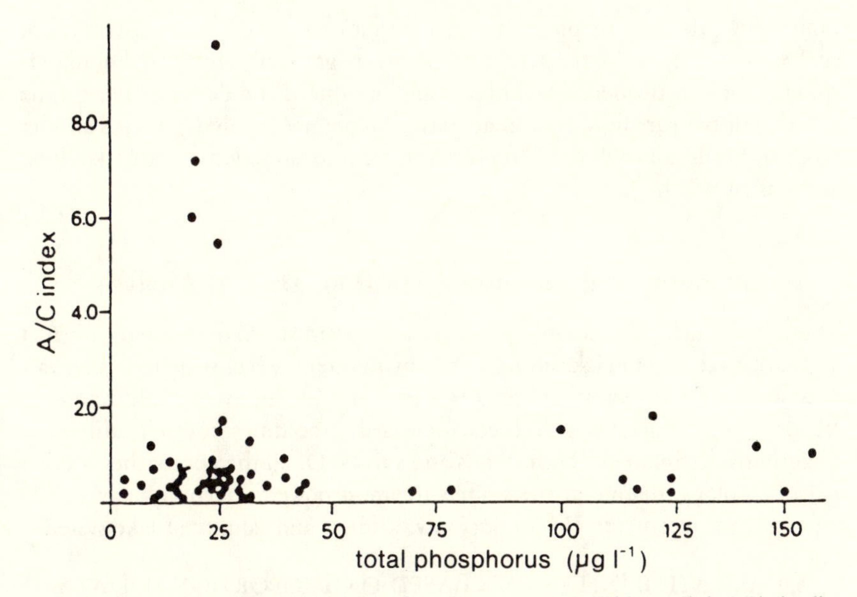

Fig. 11. A/C ratio versus total phosphorus (from Brugam, 1979). (With permission: Blackwell Scientific Publications Limited.)

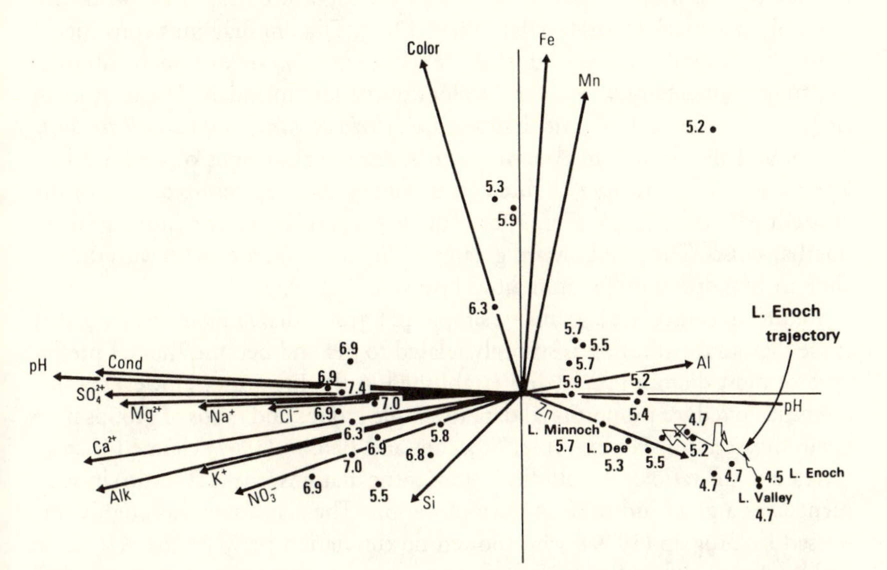

Fig. 12. Canonical correspondence analysis (CCA) time track for the Loch Enoch core (see Fig. 10) constrained by the surface sediment assemblages of the regional data-set (filled circles). The number beside each data point is the pH of the lake.

limited by problems of preservation, identification, abundance, diversity, or ecological understanding. As these problems are gradually eliminated or understood, as new instrumental techniques are introduced and developed (e.g., the use of high-pressure liquid chromatography in pigment analysis), and as a wider range of fossils are analyzed, a more complete reconstruction of past lake floras and faunas will be possible.

Environmental Reconstruction from Diatom Analysis

The aim of much paleolimnological work is to use the fossil record to reconstruct and explain changes in lake ecology and environment. Reconstructions have various levels of sophistication, ranging from the qualitative establishment of trends using subjective and literature-based procedures to well-calibrated, quantitative estimates of both trends and values. Of all the approaches used in paleolimnology, diatom analysis offers the most powerful technique available, especially for reconstructing productivity, acidity, and salinity of lake systems.

Qualitative Inferences Based on Proportional Data

The high concentration and diversity of diatoms in lake sediments allow the percentage contribution of individual taxa to the total to be expressed with considerable statistical precision (Battarbee, 1986). Diatom diagrams constructed from proportional analyses of this nature (e.g., Fig. 10a) are extremely informative to specialists familiar with the ecology of the taxa included. The increase in frequency of the acidobiontic diatoms *Tabellaria quadriseptata* and *Tabellaria binalis* and the decline in *Anomoeoneis vitrea* are clear signs of acidification. The trends can be further clarified by grouping the diatoms according to the Hustedt pH classification (Fig. 10b). The diagram then becomes intelligible to nondiatomists. The trend toward greater acidity began in the 1850s with the decline in proportion of circumneutral taxa.

Reducing complex diagrams to simple pH-group diagrams is successful because diatom distribution is strongly related to pH and because the pH preference of most diatom taxa is well established in the literature. However, some attempts to reduce proportional data to simple groups and ratios of groups have been unsuccessful. For example, Stockner and Benson (1967) claimed that an increase in the ratio of Araphidinate to Centric diatoms (the A/C ratio) in sediments was a good indicator of eutrophication. The claim was thoroughly dismissed by Brugam (1979), who showed no correlation between the A/C ratio and total phosphorus for a series of lakes across a productivity gradient in Minnesota (Fig. 11).

A more objective way of establishing trends is to compare sediment core assemblages with modern surface-sediment assemblages, that contains potentially analogous floras and for which modern water-chemical data are available. In Fig. 12 canonical correspondence analysis (ter Braak, 1986, 1987a) has been used to relate the core data from Loch Enoch (Fig. 10) directly to a series of sites in the Loch Enoch region, Galloway, S.W. Scotland. The modern sites range from pH 4.5 to pH 7.0 (Flower, 1986), and the first axis of the ordination is strongly related to pH and associated variables. It is clear from these data that Loch Enoch has always been a very acid lake. The lowermost samples in the core have diatom assemblages similar to those currently found in Loch Dee and Loch Skerrow. Nevertheless a marked acidification is shown by the trend in sample scores parallel to the first axis and toward floras similar to those at the very acid sites Loch Valley and Loch Grannoch.

QUALITATIVE INFERENCES BASED ON CONCENTRATION AND ACCUMULATION RATE DATA

In an attempt to avoid the closure problems of proportional data, some workers have used diatom concentrations and diatom accumulation rates for environmental reconstruction. In principle this approach is especially suited to eutrophication studies (Battarbee, 1973a, 1978a; Bradbury and Waddington, 1973; Moss, 1978), where it can be argued that productivity increases in lakes should cause an increase in the annual flux of diatoms to the sediment. There are no major technical problems in calculating diatom accumulation rate over the recent past. Diatom concentrations can be estimated with good precision (Battarbee, 1973b; Battarbee and Kneen, 1982), and sediment accumulation rates can be derived from ^{210}Pb dating. Differences in diatom size can be accommodated by converting data to biovolume accumulation rates (Battarbee, 1978a); problems of diatom fracture in sediments can be overcome by using biogenic silica as a surrogate for diatom biomass (Renberg, 1976; Schelske *et al.*, 1986).

Using these techniques in a study of the eutrophication of Lough Neagh, Battarbee (1978a) was able to show that a major increase in the accumulation rate of diatoms had occurred in the upper sediments of a core (Fig. 13) from about 1,880 a.c.e. These data were in agreement with the expected timescale of eutrophication for the lake based on documentary and other paleolimnological evidence. Despite the apparent success of this technique at this site, the conclusions are based on two major assumptions: that there is a simple relationship between diatom accumulation rate and lake primary productivity and that accumulation-rate data from a single core represent the mean for the lake as a whole.

The first assumption has never been tackled systematically, though it can be argued that nutrient enrichment should cause an increase in diatom production at least until the point at which dissolved silica becomes a limiting nutrient (Battarbee, 1978c). The second assumption has been questioned recently by Ander-

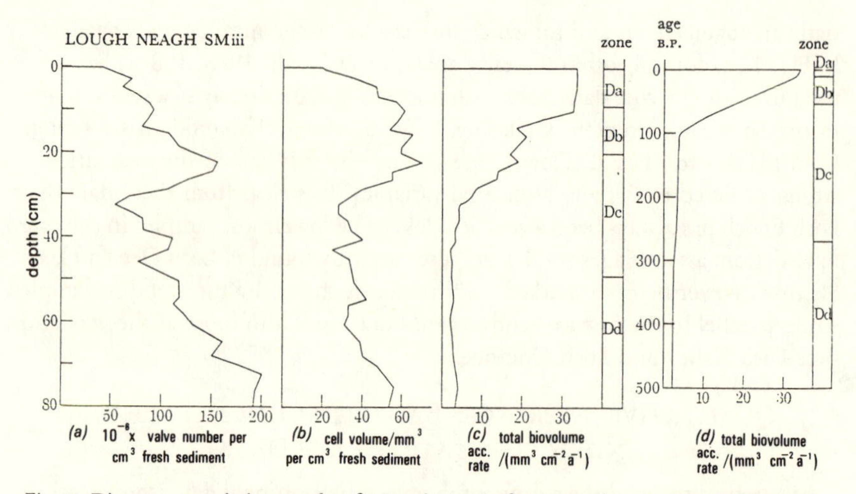

Fig. 13. Diatom accumulation rate data for Lough Neagh (from Battarbee, 1978a). (With permission: The Royal Society of London.)

son (1986b, 1989), who compared the diatom accumulation rate of a number of cores from Lough Augher, Northern Ireland, with the whole-basin mean for each diatom assemblage zone. Since the lake was known to have been strongly enriched over the last 80 years by effluent from a nearby creamery, a trend recorded in the sediments by the development of a *Stephanodiscus*-dominated phytoplankton, the site was ideally suited for this kind of evaluation.

Figure 14 shows that there has been an increase in the diatom accumulation rate in the lake sediment when data from all cores are averaged on a zone by zone basis; however, trends from individual cores vary. Not all cores show the expected increase. These differences are more clearly demonstrated when the ratios of the zone values for each core to the basin mean-zone values are compared (Fig. 15). Although AA10 and AA12 show a fairly constant relationship to the zone mean, others do not. Some underrepresent the mean, others are overrepresentative, and in many cores the pattern changes through time. Consequently it is not easy to choose in advance one or two core locations in the lake that might constantly reflect the basin mean. It is certainly not the core from the deepest point of the lake (AA4).

Because of these complexities and because of alternative possibilities, diatom accumulation rates, whether based on single cores or on basin mean values, are not the most effective way to reconstruct productivity history. However, they are essential if it is necessary to calculate lake silica budgets (Battarbee, 1973a) and are very useful in expressing species trends in lakes when it is necessary to avoid the reciprocal responses of other species that occur in proportional analyses (Battarbee, 1978c; Engstrom *et al.*, 1985; Flower *et al.*, 1987).

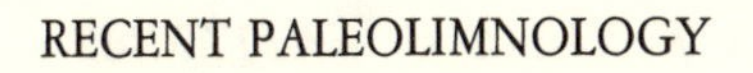

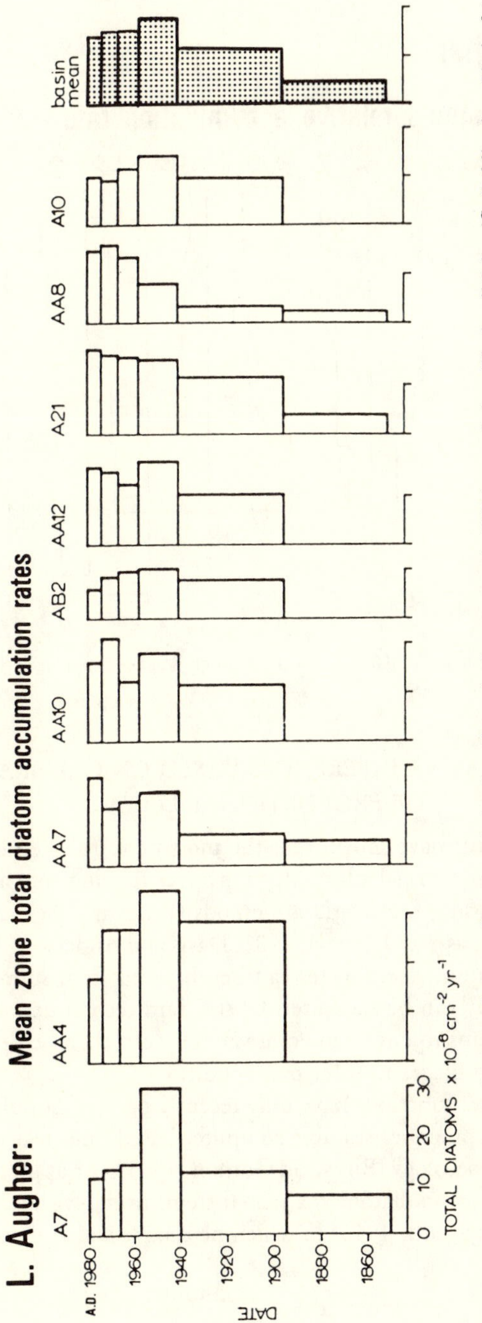

Fig. 14. Diatom accumulation rates for cores from Lough Augher (from Anderson, 1989). (With permission: Blackwell Scientific Publications Limited.)

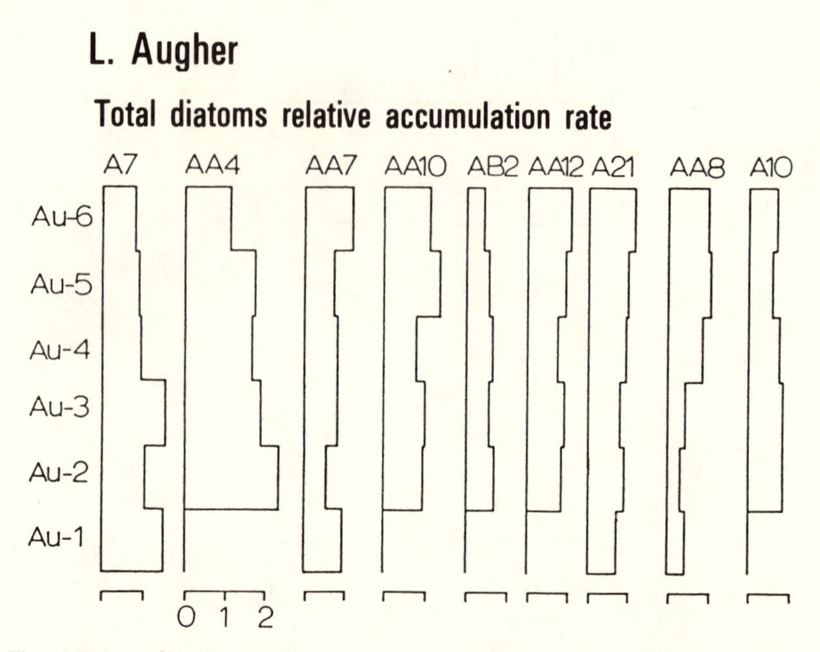

Fig. 15. Relative diatom accumulation rates from Lough Augher (from Anderson, 1986b, with author's permission).

Quantitative Inferences Based on Calibration of Proportional Data

Recent developments have shown that the most powerful method of environmental reconstruction in paleolimnology involves relating core assemblages to modern data-sets using multivariate methods (Brugam, 1983; Huttunen and Meriläinen, 1983; Gasse and Tekaia, 1983; Davis and Anderson, 1985; Charles, 1985). Modern diatom assemblages can be obtained from surface sediments, and water chemistry can be measured by standard techniques. Transfer functions relating diatoms to environmental variables can then be derived and applied to fossil assemblages to infer past conditions.

Although transfer functions have only recently become popular and mainly postdate the development of suitable computer-based, multivariate techniques in ecology and paleoecology (Birks and Gordon, 1985), the approach has a comparatively long history in diatom analysis. It stems from Nygaard's classic paper on Store Gribsø (Nygaard, 1956) in which he established and calibrated three diatom indices:

$$\text{Index } \alpha = \frac{\text{acid units}}{\text{alkaline units}}$$

$$\text{Index } \omega = \frac{\text{acid units}}{\text{number of acid species}}$$

$$\text{Index } \varepsilon = \frac{\text{acid units}}{\text{number of alkaline species}}$$

The terminology and classification followed Hustedt's earlier pH classification (Hustedt, 1937–1939). Nygaard calibrated these indices using a modern data-set of diatoms and measured pH from a range of Danish lakes. However, the application of the "transfer function" to the core assemblages was based on a subjective comparison of index scores rather than on a regression equation.

Linear Regression Analysis

Building on this early foundation Meriläinen (1967) pointed out that a regression equation that related $\log_{10}$ index α to measured water pH could be used to predict water pH to an accuracy of around one pH unit (Fig. 16). The development of this methodology by Renberg and Hellberg (1982), Davis and Anderson (1985), Charles (1985), and others is now well known and has been frequently reviewed (Battarbee, 1984; Davis, 1987). It has involved moving away from the original, simple linear-regression approach, in which the diatom index score is taken as the explanatory variable, to multiple linear regression functions, in which the explanatory variables are the individual pH preference groups. A range of equations used in the literature derived from different regional training (surface sample calibration) sets is given by ter Braak and van Dam (1989).

At the present time most pH reconstructions in acid surface waters use one or a combination of these linear regression methods. Similar techniques have been developed for pH and cladocera (Krause-Dellin and Steinberg, 1986) and for pH and a combined chrysophyte/diatom index (Charles and Smol, 1988).

Nonlinear Models

Despite the success of the linear regression approach, especially when applied to groups of species, several of the mathematical and biological assumptions of the method cannot be fully justified. Birks and Gordon (1985), with respect to pollen-climate calibrations, and Birks (1987) have listed these problems. The most important is the assumption that species have linear or monotonic relationships to environmental variables. Over short gradients and by combining species into groups (as in the pH preference method), this can be valid and problems of nonlinearity can be minimized. Nevertheless it is more appropriate to develop species-based methods that do not require prior ecological classification, especially since classification can be arbitrary. Such models can be applied robustly over long gradients.

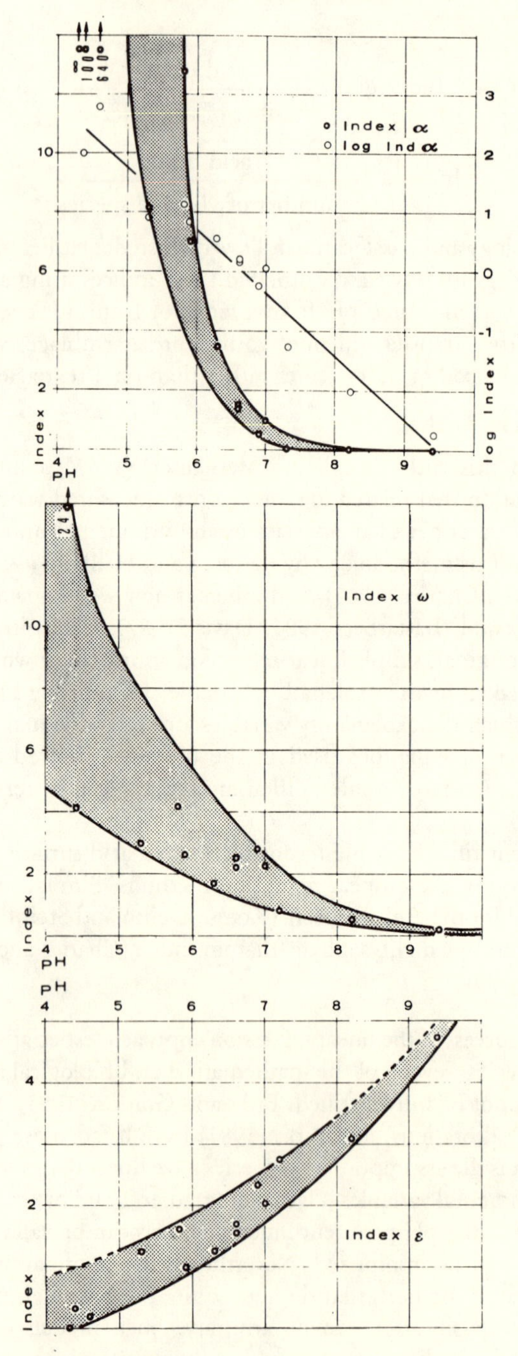

Fig. 16. Calibration of Nygaard indices (from Meriläinen, 1967). (With permission.)

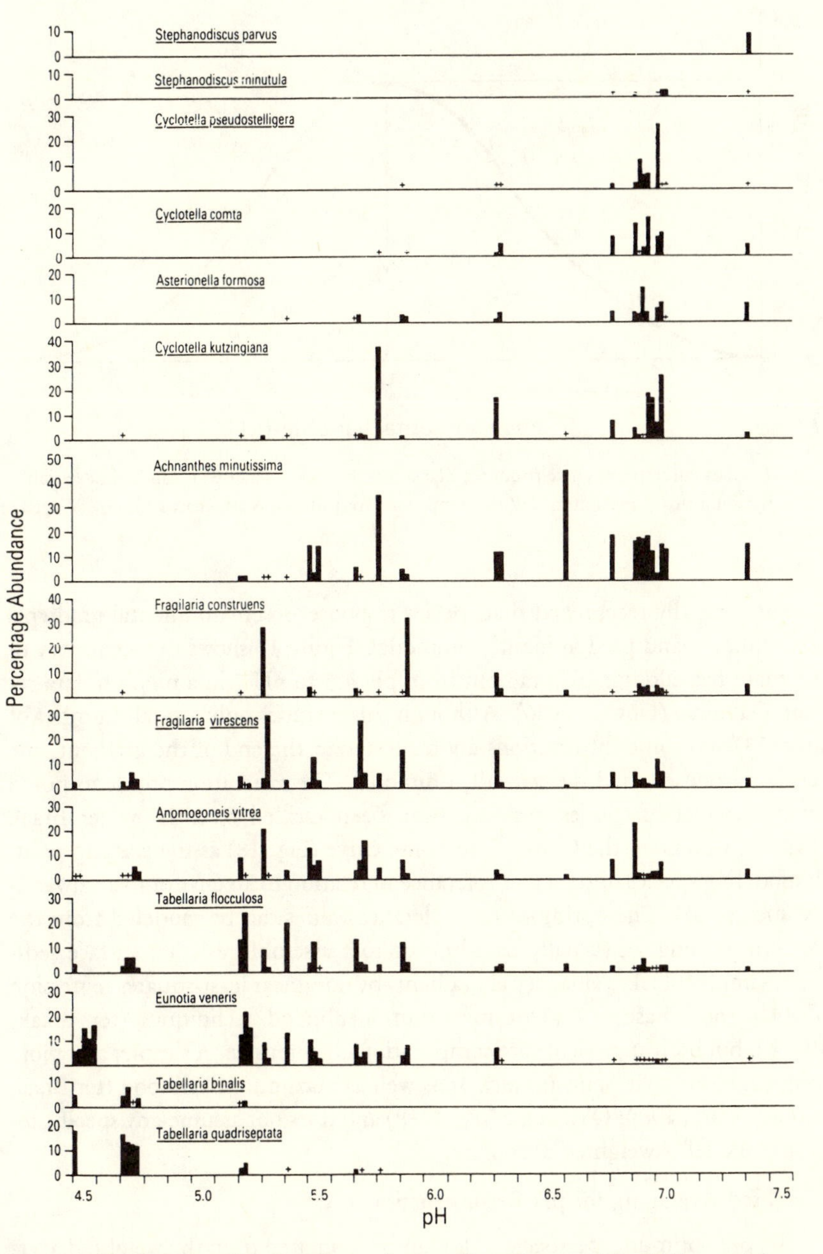

Fig. 17. Diatoms versus pH for a Galloway data-set (from Flower, 1986). (With permission: Kluwer Academic Publishers, Dordrecht.)

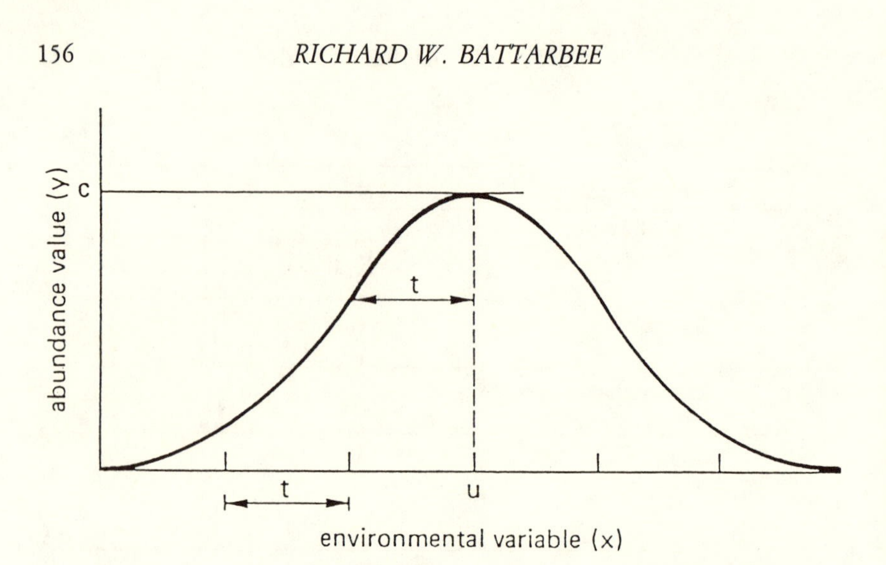

Fig. 18. Gaussian response curve (from ter Braak, 1987b). t = tolerance (1 standard deviation); u = optimum; c = maximum. (With permission: Agricultural Mathematics Group, Wageningen.)

It is generally recognized that species responses to environmental gradients are nonlinear and predominantly unimodal. Figure 17 shows the frequency of the main taxa along a pH gradient from pH 4.5 to pH 7 in a modern data-set from Galloway (Flower, 1986). Although this includes only a small number of sites (33) and some distributions are truncated at the end of the gradient, the species response to pH is essentially unimodal. The calibration problems posed by this model of species response have been tackled recently by ter Braak (1987b), who takes the Gaussian response curve (Fig. 18) as the starting point in modeling species optima and tolerance in relation to an environmental variable such as pH. The optimum and tolerance values can be modeled from the modern training set (usually based on diatom assemblages from surface sediment samples of lakes along a pH gradient) by nonlinear least-squares regression (Gauch and Chase, 1974) or maximum likelihood techniques (ter Braak, 1987b), but both approaches are computationally intensive. A simpler and more robust method that performs almost as well as maximum likelihood (ter Braak and van Dam, 1989; Oksanen *et al*., 1988) and does not assume any specific response model is weighted averaging.

Weighted Averaging for pH Reconstruction

The pH optimum of a species (U_k) can be estimated from the weighted average of the pH of lakes in which a particular species is present (ter Braak, 1987b). If the species shows a unimodal response to pH, and assuming the species occurs

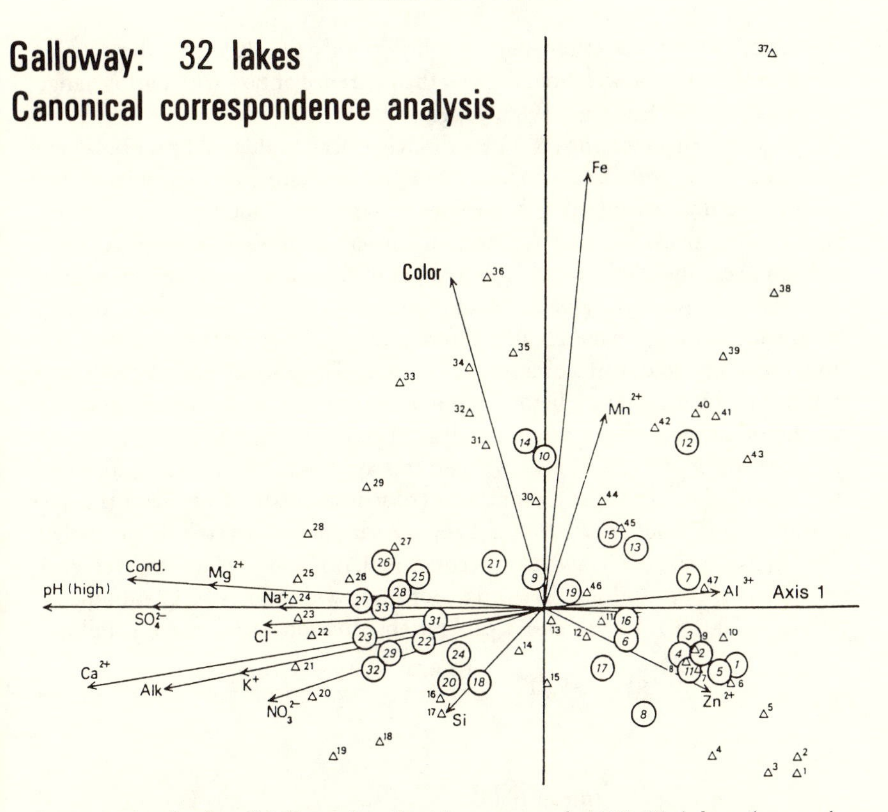

Fig. 19. CCA ordination of Galloway data (from Stevenson *et al.*, 1988). Circled numbers are the lakes sampled. Triangles show the coordinate positions of numbered species. (With permission: *Ambio*.)

most frequently and most abundantly at or near its pH optimum (ter Braak and van Dam, 1989, Stevenson *et al.*, 1988), then

$$\hat{U}_k = \frac{\Sigma y_{ik} x_i}{\Sigma y_{ik}}$$

where x_i is the value of the environmental variable x at site i, and y_{ik} is the abundance of species k at site i.

When estimates of U_k in a training data-set are accurately established, the pH of a lake can be inferred from its diatom composition, assuming that for any lake with a particular pH, taxa with an optimum close to the lake pH will be most abundant. Then

$$\hat{x}_i = \frac{\Sigma y_{ik} U_k}{\Sigma y_{ik}}$$

If necessary where species ranges or amplitudes vary greatly within the data set, tolerances can also be weighted to favor the influence of taxa with narrow ranges (ter Braak and Barendregt, 1986; Juggins, 1988).

The weighted-averaging (WA) approach has been evaluated by ter Braak and van Dam (1989) and Oksanen *et al.* (1988) by comparing its performance with existing methods in inferring pH values of samples in independent test sets. Both studies concluded that WA was an efficient technique, though ter Braak and van Dam observed that the inferred pH of the test set had a large standard error of 0.71. This high value is probably an artifact of their heterogeneous data-set in which samples were obtained from a range of microhabitats rather than from the more usual, surface sediment locations. These latter tend to give a more uniform representation of diatom populations in lakes, and—where possible—are the preferred sites for sampling (Battarbee *et al.*, 1986).

Stevenson *et al.* (1988) also evaluate this approach by comparing different methods of pH reconstruction for a series of sediment cores from acidified lakes in Galloway (Flower *et al.*, 1987). They use canonical correspondence analysis (CCA) (ter Braak, 1986, 1987a), a technique that involves weighted averaging, to ordinate species and sites in the data set along with associated environmental variables (Fig. 19). In this case a calibration equation was derived as follows:

$$\text{pH} = 5.817 + \frac{(0.824 \times \text{Axis 1 score})}{1.1}$$

where

5.817 = mean pH
0.824 = standard deviation of the measured pH in the data set
1.1 = canonical coefficient for standardized pH
Axis 1 score = score of each sample whose diatom assemblage is known but whose pH value is not known and is to be estimated

and then used as a transfer function after introducing fossil assemblages into the ordination as "passive samples." The scores of these samples on axis one were then entered into the equation.

Even with the limited data-set, this approach performs as well as other techniques for this site (Fig. 20). Problems do occur when taxa are either poorly represented or absent from the modern calibration data-set but are common in fossil assemblages and when the species ranges of common taxa are severely truncated.

These difficulties can be reduced by enlarging the modern calibration data-set as in the major collaborative lake acidification projects: Paleoecological Investigation of Recent Lake Acidification (PIRLA) in North America (Charles and

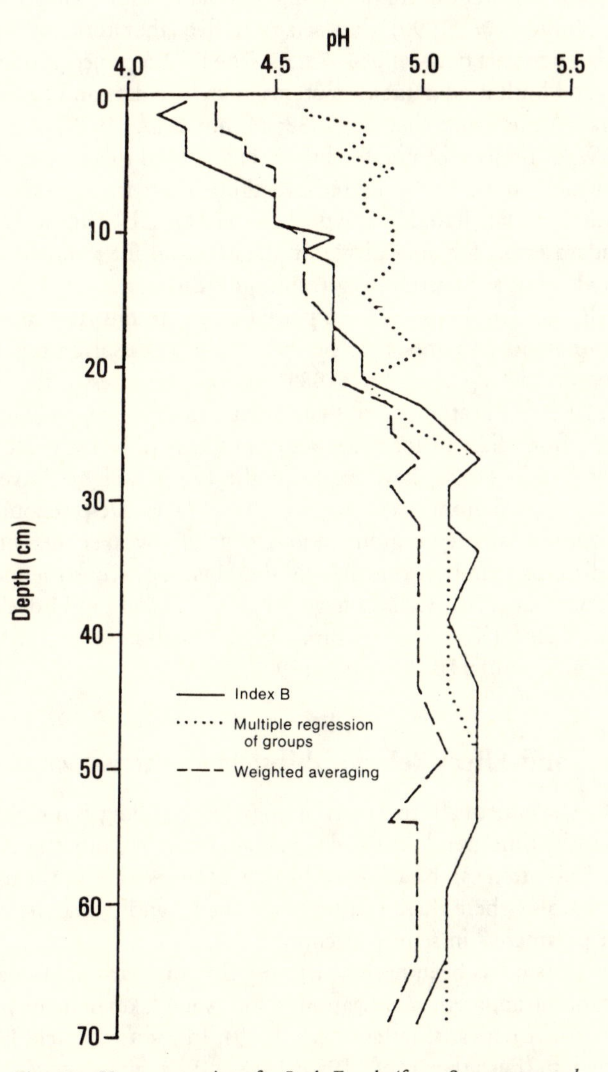

Fig. 20. pH reconstructions for Loch Enoch (from Stevenson *et al.*, 1988). Three different reconstructions are shown. Index B as defined by Renberg and Hellberg (1982), multiple regression of groups (Flower, 1986), and weighted averaging (Stevenson *et al.*, 1988). (With permission: *Ambio*.)

Whitehead, 1986; Charles *et al.*, 1989) and Surface Water Acidification Project (SWAP) in Europe (Battarbee and Renberg, 1985; Battarbee *et al.*, 1990). In the SWAP project, after careful taxonomic standardization (Kreiser and Battarbee, 1988; Munro *et al.*, 1990), data sets from five laboratories were combined to create a large project data-set of 167 sites. The PIRLA project followed similar procedures and both used database systems to archive data and to enable selective retrieval (Ahmad and Charles, 1988; Munro *et al.*, 1990).

In the SWAP project the enlarged data set was used to generate diatom-pH transfer functions using the weighted averaging approach (Birks *et al.*, 1990) with computations facilitated by the program WACALIB. This program also allowed standard errors for the calibration data-set and for reconstructed pH to be calculated using a bootstrapping technique (Birks *et al.*, 1990b).

Although most emphasis has been placed on pH reconstruction, diatoms in acidic environments also appear to be influenced by dissolved organic carbon (DOC) concentrations. Davis *et al.* (1985) have already stressed the importance of DOC in lake acidification studies and used a multiple regression technique to infer DOC from diatoms for two cores from southern Norway. More recently Kingston and Birks (1990) have explored the use of weighted averaging for reconstruction from diatom assemblages in PIRLA lakes. Despite some variation in species optima between regions, and a generally weaker relation between DOC and diatoms than between pH and diatoms, the reconstructions support the hypothesis that DOC declines have occurred parallel to pH declines in recently acidified lakes (Kingston and Birks, 1990). Analysis of the SWAP data-set has given similar results (Birks *et al.*, 1990a).

Cause-and-Effect Relationships: Hypothesis Evaluation

The aim of environmental reconstruction in paleolimnology is not only to identify and quantify time-trends but also to explain them by a process of hypothesis evaluation. This often can be achieved by careful site selection, the use of space and time controls (where these can be established), and by parallel analysis of all relevant parameters in sediment cores.

In recent years it has been necessary to use this approach to evaluate various hypotheses for the apparent acidification of soft-water lakes in many parts of Europe and North America (Charles *et al.*, 1989). In a study of acid lakes in the United Kingdom (Battarbee *et al.*, 1985, 1988a, 1989; Flower *et al.* 1987; Jones *et al.*, 1989), a number of alternatives based on arguments put forward by politicians, industrialists, and other scientists were considered:

1. The lakes may be naturally acidic and have not changed over time.
2. Acidification has occurred slowly during the postglacial period as a result of leaching and accumulation of acidic humus in catchment soils.

3. Acidification has occurred recently (since 1800) as a result of changes in the land use and land management of lake catchments including: (a) a decline in burning and grazing and (b) an increase in afforestation.

4. Acidification has occurred recently (since 1800) and is the result of an increase in acid deposition from the combustion of fossil fuels.

It is relatively trivial in most cases for a paleolimnologist to evaluate the first and second hypotheses. In sites from the United Kingdom, evidence for a marked reduction in lake-water pH is usually confined to the most recent sediments, postdating 1800 a.c.e. and usually following long periods of very stable conditions (Battarbee *et al.*, 1988a; Jones *et al.*, 1989). Hypotheses of no change can equally easily be assessed in lake eutrophication problems (Digerfeldt, 1972; Bradbury, 1975; Battarbee, 1978a, 1986; Brugam, 1978; Engstrom *et al.*, 1985).

Some aspects of the land-use hypothesis are equally easy to assess. For example, the hypothesis that lakes are acidified because of postwar afforestation in the U.K. can be disproved at some sites simply by showing that adjacent nonafforested and afforested sites have been acidified over the same timescales and that the acidification of the afforested sites occurred before planting (Flower and Battarbee, 1983; Flower *et al.*, 1987).

When the timescales of land-use change are the same as the timescale of changes in acid deposition or where land-use/management changes are poorly documented, space-time controls of this nature are more difficult to apply directly. However, it is possible to remove the influence of acid deposition either by examining a sensitive site with land-use change in a "clean" area or by examining the lake response to an analogous change in soil or vegetation in the "clean" (pre-1800 a.c.e.) past. Such a "past analog" can be identified by pollen analysis. For example, Jones *et al.* (1986) have shown that the whole-scale development of *Calluna* heathland and the initiation and spread of blanket mires in the catchment of the Round Loch of Glenhead 4,000–6,000 years ago (Fig. 21) did not cause the pH of the lake to fall below about pH 5.5, whereas a rapid decline in the pH of this lake did occur after 1850 a.c.e. coincident with the rise in SO_2 emissions in the U.K. (Barrett *et al.*, 1983; Renberg *et al.*, 1990). Anderson and Korsman (1990) have used a similar approach. Renberg evaluated the impact of recent *Picea* (spruce) expansion on acidification and found by diatom analysis of sediments that no change in water quality resulted from the natural immigration of spruce 3,000 years ago.

These data show that land-use changes in the absence of acid deposition have little effect on lake acidity; however, land-use changes in association with acid deposition can be important. Modern data from adjacent afforested and moorland streams in Scotland (Harriman and Morrison, 1982) showed that water draining from afforested catchments was more acidic than water from the moor-

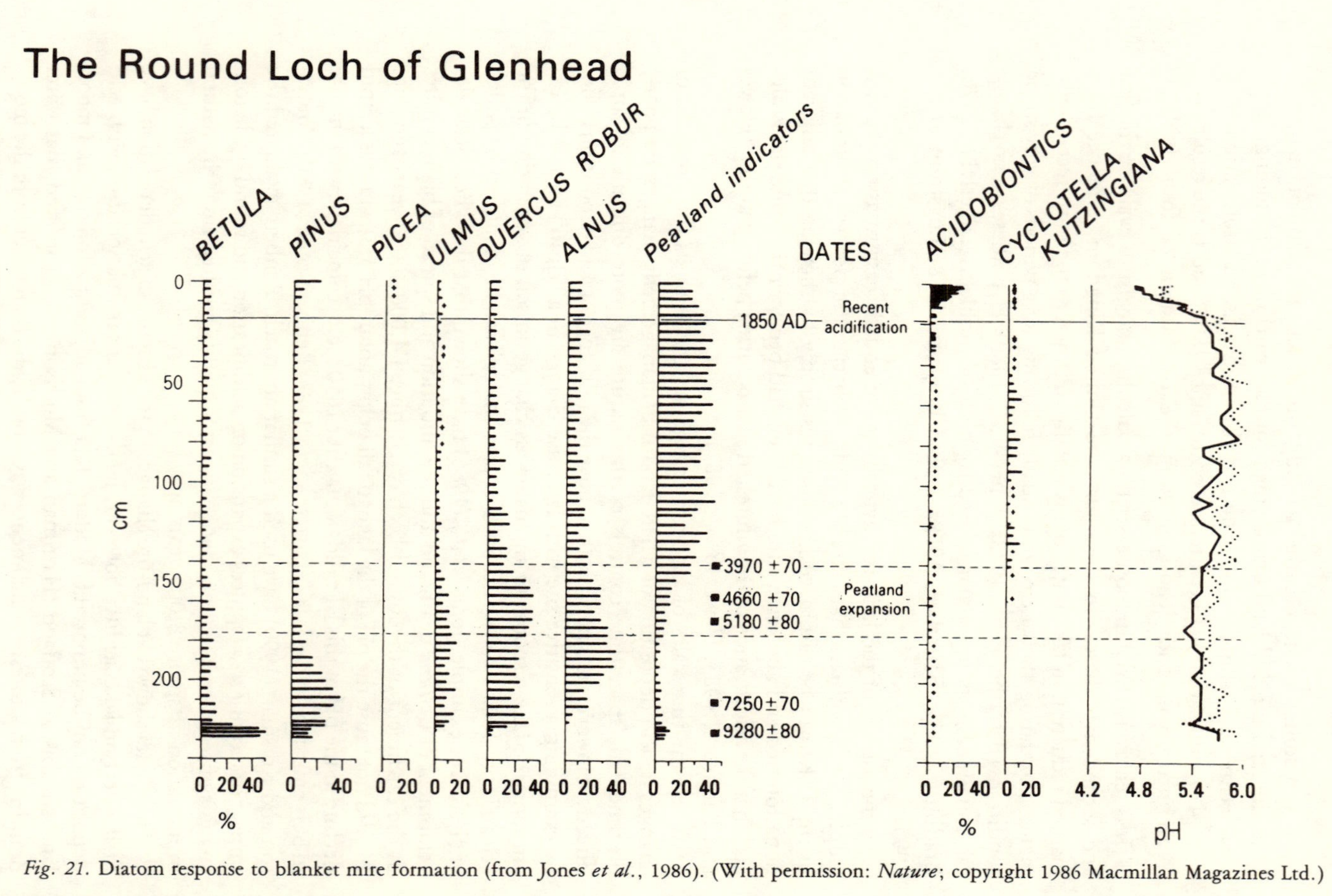

Fig. 21. Diatom response to blanket mire formation (from Jones *et al.*, 1986). (With permission: *Nature*; copyright 1986 Macmillan Magazines Ltd.)

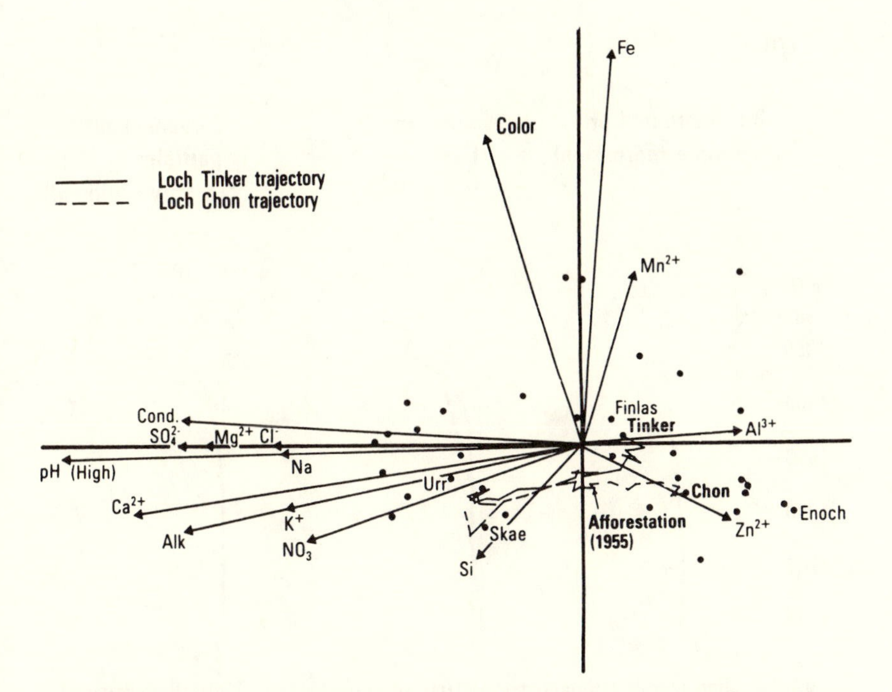

Fig. 22. CCA time tracks of cores from Loch Chon and Loch Tinker constrained by the regional data-set (Kreiser and Battarbee, unpublished).

land controls. Harriman and Morrison (1982) reasoned that this "forest effect" was a result of more effective pollution scavenging of the atmosphere by the forest than by open moorland.

Because previous studies used sensitive sites that were already strongly acidified before forest planting (Flower *et al.*, 1987, and above), Kreiser *et al.* (1990) chose to examine a pair of adjacent and somewhat less-sensitive lake sites to evaluate this effect, Loch Chon and Loch Tinker. The present pH of Loch Chon, the afforested site, is 5.2 whereas the nonafforested site, Loch Tinker, has a pH of 6.0. Since both have very similar Ca^{++} values (about 80 $\mu eq\ l^{-1}$), they should have similar pH values unless the difference is a result of the effect of the forest or of some unknown geochemical difference in the respective catchments. If the latter case were true, Loch Chon would be expected to have had a more acidic diatom flora and lower pH throughout its history.

Figure 22 shows the comparative time-tracks of short cores from the two sites in the context of the Galloway modern data-set (cf. Fig. 19 above). Both sites had very similar floras in the mid-19th century, with reconstructed pH values of about 6.4, indicating no major geochemical difference between the two sites. In the late 19th and early 20th centuries, before afforestation, both cores track

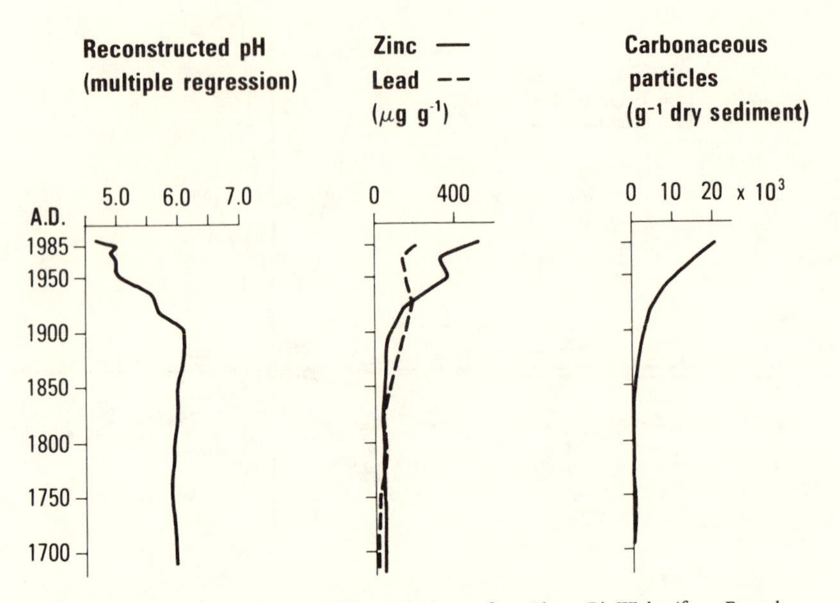

Fig. 23. Sediment record of atmospheric contaminants from Llyn y Bi, Wales (from Battarbee *et al.*, 1988a). (With permission: ENSIS Ltd., London.)

parallel to the first axis toward acidification. After afforestation in the 1950s, however, the Loch Chon trajectory continues in the direction of acidification toward floras typical of some of the most acidic sites in the modern data-set. The Loch Tinker flora tends to stabilize and then reverse to somewhat-less-acidic conditions. Reversals in acidity have been observed at other moorland sites in Scotland (Battarbee *et al.*, 1988b), probably because of a 30–40% reduction in acid deposition since the early 1970s (Barrett *et al.*, 1987). The trend at Loch Chon, on the other hand, is consistent with the forest effect, and the lack of a reversal at this site suggests the effect is strong enough to counter the reduction in acid deposition.

Since this forest effect is less pronounced in low-acid-deposition areas (Kreiser *et al.*, 1990) or at times in the past when acid deposition can be assumed to be minimal (Renberg *et al.*, 1990), none of the alternative hypotheses have paleolimnological support. On the other hand there is considerable evidence from lake sediments that the acid-deposition hypothesis is the most plausible (Battarbee *et al.*, 1988a).

At all acidified sites studied so far in the U.K., the following observations can be made: (1) The beginning of acidification postdates 1,800 a.c.e. (2) Acidification never begins before evidence of trace metal contamination. (3) All sites have high concentrations of carbonaceous particles that are derived from fossil-fuel combustion (Fig. 23). Moreover, allowing for variations in catchment sensitivity, a simple dose-response model between S deposition and lake acidification clearly predicts the distinctions between acidified and nonacidified lakes (Battarbee, 1990).

Conclusions

Interest in lake pollution, especially acidification, has led to a marked increase in paleolimnological research over recent decades and to the development of improved techniques for coring and dating recent lake sediments. In addition the range of analytical techniques available for biological and environmental reconstruction has been steadily expanding, and appropriate statistical, computer-based techniques are now available for data analysis. Many research groups are firmly committed to the construction of large sets of modern data that can be used to derive transfer functions, not only for pH but for a range of other environmental variables. Such data sets can be expanded into eutrophic and saline lake systems to derive transfer functions for productivity (Deevey *et al.*, 1986; Whitmore, 1989) and salinity (Radle *et al.*, 1989; Fritz, 1990; Fritz and Battarbee, 1988), which ultimately can be applied to paleolimnological questions on all timescales including lake ontogeny (Deevey, 1984) and climatic change (Winter and Wright, 1977). In conjunction with other lake-sediment-based techniques and with careful research design, recent paleolimnology can help to explain environmental change in drainage-basin systems in the integrated and comprehensive way envisaged by Oldfield (1977).

Acknowledgments

I would like to thank John Anderson, Roger Flower, Tony Stevenson, Viv Jones, Steve Juggins, Martin Munro, Annette Kreiser, Peter Appleby, Frank Oldfield, John Smol, Sheri Fritz, Dan Engstrom, John Birks, John Kingston, Brian Rippey, Ingemar Renberg, Gunnar Digerfeldt, Jouko Meriläinen, Pertti Huttunen, Heikki Simola, and Herb Wright for their direct and indirect, knowing and unknowing help with this paper. I am especially grateful to John Birks and Tony Stevenson for providing data analyses of various kinds, some of which have been included as figures. Herb Wright has given me much valued assistance, support, and friendship over the years, and it gives me great pleasure to dedicate this chapter to him.

References

Ahmad, H., and Charles, D. F. (1988). "PIRLA Data Base Management System User's Manual" (2nd ed.). Palaeoecological Investigation of Recent Lake Acidification (PIRLA) Report no. 32.

Anderson, N. J. (1986a). Diatom biostratigraphy and core correlation within a small lake basin. *Hydrobiologia* **143**, 105–112.

——. (1986b). "Recent sediment accumulation in a small lake basin with special reference to diatoms." Unpublished Ph.D. thesis, University of London.

——. (1989). A whole basin diatom accumulation rate for a small eutrophic lake in Northern Ireland and its ecological implications. *Journal of Ecology* 77, 926–946.

Anderson, N. J., Battarbee, R. W., Appleby, P. G., Stevenson, A. C., Oldfield, F., Darley, J., and Glover, G. (1986). Palaeolimnological evidence for the recent acidification of Loch Fleet, Galloway. Research Paper No. 17, Palaeoecology Research Unit, University College London.

Anderson, N. J., and Korsman, T. (1990). Land-use change and lake acidification: Iron age desettlement in northern Sweden as a pre-industrial analogue. *Philosophical Transactions of the Royal Society, London B* **327**, 373–376.

Appleby, P. G., Nolan, P. J., Gifford, D. W., Godfrey, M. J., Oldfield, F., Anderson, N. J., and Battarbee, R. W. (1986). ^{210}Pb dating by low background gamma counting. *Hydrobiologia* **143**, 21–27.

Appleby, P. G., and Oldfield, F. (1978). The calculation of lead-210 dates assuming a constant rate of supply of unsupported ^{210}Pb to the sediment. *Catena* **5**, 1–8.

——. (1983). The assessment of ^{210}Pb data from sites with varying sediment accumulation rates. *Hydrobiologia* **103**, 29–35.

——. (1988). Radioisotope studies of recent lake and reservoir sedimentation. *In* "The Use of Nuclear Techniques in Sediment Transport and Sedimentation Problems" (M. J. Crickmore *et al.*, Eds.). UNESCO.

Barrett, C. F., Atkins, D. H. F., Cape, J. N., Crabtree, J., Davies, T. D., Derwent, R. G., Fisher, B. E. A., Fowler, D., Kallend, A. S., Martin, A., Scriven, R. A., and Irwin, J. G. (1987). "Acid Deposition in the United Kingdom 1981–1985: A Second Report of the United Kingdom Review Group on Acid Rain." Warren Spring Laboratory, Stevenage, U.K.

Barrett, C. F., Atkins, D. H. F., Cape, J. N., Fowler, D., Irwin, J. G., Kallend, A. S., Martin, A., Pitman, J. I., Scriven, R. A., Tuck, A. F., and Irwin, J. G. (1983). "Acid Deposition in the United Kingdom: Report of the United Kingdom Review Group on Acid Rain." Warren Spring Laboratory, Stevenage, U.K.

Battarbee, R. W. (1973a). Preliminary studies of Lough Neagh sediments II. Diatoms from the uppermost sediment. *In* "Quaternary Plant Ecology" (H. J. B. Birks and R. G. West, Eds.), pp. 279–289. Blackwell, Oxford.

——. (1973b). A new method for estimating absolute microfossil numbers with special reference to diatoms. *Limnology and Oceanography* **18**, 647–653.

——. (1978a). Observations on the recent history of Lough Neagh and its drainage basin. *Philosophical Transactions of the Royal Society, London* B **281**, 303–345.

——. (1978b). Biostratigraphical evidence for variations in the recent pattern of sediment accumulation in Lough Neagh, Northern Ireland. *Verhandlungen der Internationale Vereinigung für theoretische und angewandte Limnologie* **20**, 625–629.

——. (1978c) Relative composition, concentration and calculated influx of diatoms from a sediment core from Lough Erne, Northern Ireland. *Polskie Archiv Hydrobiologie* **25**, 9–16.

——. (1979). Early algological records—help or hindrance to palaeolimnology? *Nova Hedwigia Beiheft* **64**, 379–394.

——. (1981a). Diatom and Chrysophyceae in a small meromictic lake. *In* "Florilegium Florinis Dedicatum" (L-K. Königsson and K. Paabo, Eds.), pp. 105–109. Uppsala.

——. (1981b). Changes in the diatom microflora of a eutrophic lake since 1900 from a comparison of old algal samples and the sedimentary record. *Holarctic Ecology* **4**, 73–81.

——. (1984). Diatom analysis and the acidification of lakes. *Philosophical Transactions of the Royal Society, London* B **305**, 451–477.

——. (1986). Diatom analysis. *In* "Handbook of Holocene Palaeoecology and Palaeohydrology" (B. E. Berglund, Ed.), pp. 527–570. Wiley, Chichester, England.

——. (1990). The causes of lake acidification with special reference to the role of acid deposition. *Philosophical Transactions of the Royal Society, London* B **327**, 339–347.

Battarbee, R. W., Anderson, N. J., Appleby, P. G., Flower, R. J., Fritz, S. C., Haworth, E. Y., Higgitt, S., Jones, V. J., Kreiser, A., Munro, M. A. R., Natkanski, J., Oldfield, F., Patrick, S. T., Richardson, N. G., Rippey, B., and Stevenson, A. C. (1988a). Lake acidification in the United Kingdom 1800–1986: Evidence from analysis of lake sediments. Ensis, London. 68 pp.

Battarbee, R. W., Cronberg, G., and Lowry, S. (1980). Observations on the occurrence of scales and bristles of *Mallomonas* spp. (Chrysophyceae) in the micro-laminated sediments of a small lake in Finnish North Karelia. *Hydrobiologia* **71**, 225–232.

Battarbee, R. W., Flower, R. J., Stevenson, A. C., Jones, V. J., Harriman, R., and Appleby, P. G. (1988b). Diatom and chemical evidence for reversibility of acidification of Scottish lochs. *Nature* **322**, 530–532.

Battarbee, R. W., Flower, R. J., Stevenson, A. C., and Rippey, B. (1985). Lake acidification in Galloway: A palaeoecological test of competing hypotheses. *Nature* **314**, 350–352.

Battarbee, R. W., and Kneen, M. J. (1982). The use of electronically counted microspheres in absolute diatom analysis. *Limnology and Oceanography* **27**, 184–188.

Battarbee, R. W., Mason, J., Talling, J. F., and Renberg, I., Eds. (1990). "Palaeolimnology and Lake Acidification." *The Royal Society of London*.

Battarbee, R. W., and Renberg, I. (1985). Royal Society Surface Water Acidification Project (SWAP), Palaeoliminology Programme. Researach Paper 12, Palaeoecology Research Unit, University College London.

Battarbee, R. W., Smol, J. P., and Meriläinen, J. (1986). Diatoms as indicators of pH: A historical review. *In* "Diatoms and Lake Acidity: The Use of Siliceous Algal Microfossils in Reconstructing pH" (J. P. Smol, R. W. Battarbee, R. B. Davis, and J. Meriläinen, Eds.), pp. 5–14. Junk, Dordrecht.

Battarbee, R. W., Stevenson, A. C., Rippy, B., Fletcher, C., Natkanski, J., Wik, M., and Flower, R. J. (1989). Causes of lake acidification in Galloway, south-west Scotland: A palaeoecological evaluation of the relative rates of atmospheric contamination and catchment change for two acidified sites with non-afforested catchments. *Journal of Ecology* **77**, 651–672.

Birks, H. H. (1973). Modern macrofossil assemblages in lake sediments in Minnesota. *In* "Quaternary Plant Ecology" (H. J. B. Birks and R. G. West, Eds.), pp. 173–189. Blackwell, Oxford.

Birks, H. J. B. (1987). Methods for pH-calibration and reconstruction from palaeolimnological data: Procedures, problems, potential techniques. Surface Water Acidification Programme, Mid-term Review Conference, Bergen, Norway, pp. 370–380.

Birks, H. J. B., and Gordon, A. D. (1985). "Numerical Methods in Quaternary Pollen Analysis." Academic Press, London.

Birks, H. J. B., Juggins, S., and Line, J. M. (1990a). Lake Surface-water chemistry reconstructions from paleolimnological data. *In:* "The Surface Waters Acidification Program" (B. J. Mason, Ed.), pp. 301–313. Cambridge University Press, Cambridge.

Birks, H. J. B., Line, J. M., Juggins, S., Stevenson, A.C., and ter Braak, C. J. F. (1990b). Diatoms and pH reconstruction. *Philosophical Transactions of the Royal Society, London* B **327**, 263–278.

Bloemendal, J., Oldfield, F., and Thompson, R. (1979). Magnetic measurements used to assess sediment influx at Llyn Goddionduon. *Nature* **280**, 50–53.

Bonny, A. P., and Allen, P. V. (1984). Pollen recruitment to the sediments of an enclosed lake in Shropshire, England. *In* "Lake Sediments and Environmental History" (E. Y. Haworth and J. W. G. Lund, Eds.), pp. 231–260. Leicester University Press.

Bradbury, J. P. (1975). Diatom stratigraphy and human settlement in Minnesota. Geological Society of America, Special Paper 171, 74 pp.

Bradbury, J. P., Leyden, B., Salgado-Labouriau, M., Lewis, W. M., Jr., Schubert, C., Binford, M. W., Frey, D. G., Whitehead, D. R., and Weibezahn, F. H. (1981). Late Quaternary environmental history of Lake Valencia, Venezuela. *Science* **214**, 1299–1305.

Bradbury, J. P., and Waddington, J. C. B. (1973). The impact of European settlement on Shagawa Lake, Northeastern Minnesota, USA. *In* "Quaternary Plant Ecology" (H. J. B. Birks and R. G. West, Eds.), pp. 289–307. Blackwell, Oxford.

Brugam, R. B. (1978). The human disturbance history of Linsley Pond, North Branford, Connecticut. *Ecology* **59**, 19–36.

——. (1979). A re-evaluation of the Araphidineae/Centrales index as an indicator of lake trophic status. *Freshwater Biology* **9**, 451–460.

——. (1983). The relationship between fossil diatom assemblages and limnological conditions. *Hydrobiologia* **98**, 223–235.

Charles, D. F. (1985). Relationships between surface sediment diatom assemblages and lakewater characteristics in Adirondack lakes. *Ecology* **66**, 994–1011.

Charles, D. F., Battarbee, R. W., Renberg, I., van Dam, H., and Smol, J. P. (1989). Palaeoecological analysis of lake acidification trends in North America and Europe using diatoms and chrysophytes. *In* "Advances in Environmental Sciences." V. 4. "Soils, Aquatic Processes, and Lake Acidification" (S. A. Norton, S. E. Lindberg, and A. L. Page, Eds.), pp. 207–276. Springer-Verlag, New York.

Charles, D. F., and Norton, S. A. (1986). Paleolimnological evidence for trends in atmospheric deposition of acids and heavy metals. *In* "Acid Deposition: Long-term Trends," pp. 335–431. National Academy Press, Washington, D.C.

Charles, D. F., and Smol, J. P. (1988). New methods for using diatoms and chrysophytes to infer past pH of low-alkalinity lakes. *Limnology and Oceanography* **33**, 1451–1462.

Charles, D. F., and Whitehead, D. R. (1986). The PIRLA Project: Paleoecological investigations of recent lake acidification. *Hydrobiologia* **143**, 13–20.

Chivas, A. R., de Deckker, P., and Shelley, J. M. G. (1986). Magnesium and strontium in non-marine ostracod shells as indicators of palaeosalinity and palaeotemperature. *Hydrobiologia* **143**, 135–142.

Crisman, T. L., Crisman, U. A. M., and Binford, M. W. (1986). Interpretation of bryozoan microfossils in lacustrine sediment cores. *Hydrobiologia* **143**, 113–118.

Cronberg, G. (1973). Development of cysts in Mallomonas (Chrysophyceae) studied by scanning electron microscopy. *Hydrobiologia* **43**, 29–38.

——. (1982). Phytoplankton changes in Lake Trummen induced by restoration. *Folia Limnologica Scandinavica* **18**, 1–119.

——. (1986). Blue-green algae, green algae and Chrysophyceae in sediments. *In* "Handbook of Holocene Palaeoecology and Palaeohydrology" (B. E. Berglund, Ed.), pp. 507–526. Wiley, Chichester, England.

Cronberg, G., and Sandgren, C. D. (1986). A proposal for the development of standardized nomenclature and descriptive terminology for future publications concerning chrysophycean statospores. *In* "Chrysophytes—Aspects and Problems" (J. Kristiansen and R. A. Andersen, Eds.), pp. 317–328. Cambridge University Press, Cambridge.

Dakin, W. J., and Latarche, M. (1913). The plankton of Lough Neagh. *Proceedings of the Royal Irish Academy* B **30**, 20–96.

Davis, R. B. (1987). Paleolimnological diatom studies of acidification of lakes by acid rain: an application of Quaternary science. *Quaternary Science Reviews* **6**, 147–163.

Davis, R. B., and Anderson, D. S. (1985). Methods for pH calibration of sedimentary diatom remains for reconstructing history of pH in lakes. *Hydrobiologia* **120**, 69–87.

Davis, R. B., Anderson, D. S., and Berge, F. (1985). Loss of organic matter, a fundamental process in lake acidification: Palaeolimnological evidence. *Nature* **316**, 436–438.

Davis, R. B., Hess, C. T., Norton, S. A., Hanson, D. W., Hoagland, K. D., and Anderson, D. W. (1984). ^{137}Cs and ^{210}Pb dating of sediments from softwater lakes in New England (USA) and Scandinavia, a failure of ^{137}Cs dating. *Chemical Geology* **44**, 151–185.

Dearing, J. A. (1983). Changing patterns of sediment accumulation in a small lake in Scania, southern Sweden. *Hydrobiologia* **103**, 59–64.

——. (1986). Core correlation and total sediment influx. *In* "Handbook of Holocene Palaeoecology and Palaeohydrology" (B. E. Berglund, Ed.), pp. 247–272. Wiley, Chichester, England.

Dearing, J. A., Elner, J. K., and Happey-Wood, C. M. (1981). Recent sediment flux and erosional processes in a Welsh upland lake-catchment based on magnetic susceptibility measurements. *Quaternary Research* **16**, 356–372.

Deevey, E. S. (1942). Studies on Connecticut lake sediments. III. The biostratonomy of Linsley Pond. *American Journal of Science* **240**, 233–264; 313–338.

——. (1955). The obliteration of the hypolimnion. *Memorie Dell'Istituto Italiano Di Idrobiologia suppl.* **8**, 9–38.

——. (1984). Stress, strain, and stability of lacustrine ecosystems. *In* "Lake Sediments and Environmental History" (E. Y. Haworth and J. W. G. Lund, Eds.), pp. 203–229. Leicester University Press.

Deevey, E. S., Binford, M. W., Brenner, M., and Whitmore, T. J. (1986). Sedimentary records of accelerated nutrient loading in Florida lakes. *Hydrobiologia* **143**, 49–54.

Deevey, E. S., Rice, D. S., Rice, P. M., Vaughan, H. H., Brenner, M., and Flannery, M. S. (1979). Mayan urbanism: Impact on a tropical Karst environment. *Science* **206**, 298–306.

Digerfeldt, G. (1972). The post-glacial development of Lake Trummen: regional vegetation history, water-level changes and palaeolimnology. *Folia Limnologica Scandinavica* **16**, 1–96.

——. (1986). Studies on past lake-level fluctuations. *In* "Handbook of Holocene Palaeoecology and Palaeohydrology" (B. E. Berglund, Ed.), pp. 127–144. Wiley, Chichester, England.

Digerfeldt, G., Battarbee, R.W., and Bengtsson, L. (1975). Report on annually laminated sediment in Lake Järlasjön, Nacka, Stockholm. *Geologiska Föreningens i Stockholm Förhandlingar* **97**, 29–40.

Douglas, M. S. V., and Smol, J. P. (1987). Siliceous protozoan plates in lake sediments. *Hydrobiologia* **154**, 13–23.

Eakins, J. D., and Morrison, R. T. (1978). A new procedure for the determination of lead-210 in lake and marine sediments. *International Journal of Applied Radioactive Isotopes* **29**, 531–536.

Engstrom, D. R., and Nelson, S. (1991). Paleosalinity from trace metals in fossil ostracods compared with observational records at Devil's Lake, N. Dakota. Palaeogeography, Palaeoclimatology, Palaeoecology **83**, 295–312.

Engstrom, D. R., and Swain, E. B. (1986). The chemistry of lake sediments in time and space. *Hydrobiologia* **143**, 37–44.

Engstrom, D. R., Swain, E. B., and Kingston, J. C. (1985). A paleolimnological record of human disturbance from Harvey's Lake, Vermont: Geochemistry, pigments and diatoms. *Freshwater Biology* **15**, 261–288.

Evans, R. D., and Rigler, F. H. (1980). Measurement of whole-lake sediment accumulation and phosphorus retention using lead-210 dating. *Canadian Journal of Fisheries and Aquatic Sciences* **37**, 817–822.

Flower, R. J. (1986). The relationship between surface sediment diatom assemblages and pH in 33 alloway lakes: Some regression models for reconstructing pH and their application to sediment cores. *Hydrobiologia* **143**, 93–104.

Flower, R. J., and Battarbee, R.W. (1983). Diatom evidence for recent acidification of two Scottish lochs. *Nature* **305**, 130–133.

Flower, R. J., Battarbee, R. W., and Appleby, P. G. (1987). The recent palaeolimnology of six acid lakes in Galloway, south-west Scotland. Diatom analysis, pH trends, and the role of afforestation. *Journal of Ecology* **75**, 797–824.

Fogg, G. E., and Belcher, J. H. (1961). Pigments from the bottom deposits of an English lake. *New Phytologist* **60**, 129–142.

Frey, D. G. (1964). Remains of animals in Quaternary lake and bog sediments and their interpretation. *Archiv für Hydrobiologie Beiheft Ergebnisse Limnologie* **2**, 1–114.

Fritz, S. C. (1989). Lake ontogeny and limnological response to prehistoric and historic land-use in Diss, Norfolk, England. *Journal of Ecology* **77**, 182–202.

——. (1990). Twentieth-century salinity and water-level fluctuations in Devil's Lake, N.D.: Test of a diatom-based transfer function. *Limnology and Oceanography* **35**, 1771–1781.

Fritz, S. C., and Battarbee, R. W. (1988). Sedimentary diatom assemblages in freshwater and saline lakes of the Northern Great Plains, North America: Preliminary results. *In* "9th International Symposium on Living and Fossil Diatoms" (F. E. Round, Ed.), pp. 265–271. J. Cramer, Stuttgart.

Gasse, F., and Tekaia, F. (1983). Transfer functions for estimating palaeoecological conditions (pH) from East African diatoms. *Hydrobiologia* **103**, 85–90.

Gauch, H. G., and Chase, G. B. (1974). Fitting the Gaussian curve to ecological data. *Ecology* **55**, 1377–1381.

Goldberg, E. D. (1963). Geochronology with lead^{-210}. *In:* "Radioactive dating," pp. 121–131. Proceedings of the symposium on radioactive dating, International Atomic Energy Agency, Vienna.

Gorham, E., and Sanger, J. E. (1972). Fossil pigments in the surface sediments of a meromictic lake. *Limnology and Oceanography* **17**, 618–622.

Griffin, J. J., and Goldberg, E. D. (1981). Sphericity as a characteristic of solids from fossil fuel burning in a Lake Michigan sediment. *Geochimica et Cosmochimica Acta* **45**, 763–769.

——. (1983). Impact of fossil fuel combustion on sediments of Lake Michigan: a reprise. *Environmental Science and Technology* **17**, 244–245.

Griffiths, M. (1978). Specific blue-green algal carotenoids in sediments of Esthwaite water. *Limnology and Oceanography* **23**, 777–784.

Harriman, R., and Morrison, B. R. S. (1982). The ecology of streams draining forested and non-forested catchments in an area of central Scotland subject to acid precipitation. *Hydrobiologia* **88**, 251–263.

Hartmann, H., and Steinberg, C. (1986). Mallomonadacean (Chrysophyceae) scales: Early biotic paleoindicators of lake acidification. *Hydrobiologia* **143**, 87–92.

Haworth, E. Y. (1980). Comparison of continuous phytoplankton records with the diatom stratigraphy in the recent sediments of Blelham Tarn. *Limnology and Oceanography* **25**, 1093–1103.

Hustedt, F. (1937–1939). Systematische und okologische Untersuchungen über den Diatomeen-Flora von Java, Bali, Sumatra. *Archiv für Hydrobiologie* (Supplement) 15 and 16.

Hutchinson, G. E., and Wollack, A. (1940). Studies on Connecticut lake sediments. II. Chemical analysis of a core from Linsley pond, North Branford. *American Journal of Science* **238**, 493–517.

Huttunen, P., and Meriläinen, J. (1978). New freezing device providing large, unmixed samples from lakes. *Annales Botanici Fennici* **15**, 128–130.

——. (1983). Interpretation of lake quality from contemporary diatom assemblages. *Hydrobiologia* **103**, 91–98.

Huttunen, P., Meriläinen, J., and Tolonen, K. (1978). The history of a small dystrophied forest lake, southern Finland. *Polskie Archiv Hydrobiologie* **25**, 189–202.

Jaakkola, T., Tolonen, K., Huttunen, P., and Leskinen, S. (1983). The use of fallout ^{137}Cs and $^{239,\,240}$Pu for dating lake sediments. *Hydrobiologia* **103**, 15–20.

Jones, V. J., and Flower, R. J. (1986). Spatial and temporal variations in periphyton communities: Palaeoecological significance. *In* "Diatoms and Lake Acidity" (J. P. Smol, R. W. Battarbee, R. B. Davis, and J. Meriläinen, Eds.), pp 87–96. Junk, Dordrecht.

Jones, V. J., Stevenson, A. C., and Battarbee, R. W. (1986). Lake acidification and the "land-use" hypothesis: A mid-post-glacial analogue. *Nature* **322**, 157–158.

——. (1989). The acidification of lakes in Galloway, south-west Scotland: A diatom and pollen study of the post-glacial history of the Round Loch of Glenhead. *Journal of Ecology* **77**, 1–23.

Juggins, S. (1988). "A diatom/salinity transfer function for the Thames estuary and its application to waterfront archaeology." Unpublished Ph.D. thesis, University of London.

Kingston, J. C., and Birks, H. J. B. (1990). Dissolved organic carbon reconstructed from diatom assemblages in PIRLA project lakes, North America. *Philosophical Transactions of the Royal Society, London* B **327**, 279–288.

Krause-Dellin, D., and Steinberg, C. (1986). Cladoceran remains as indicators of lake acidification. *Hydrobiologia* **143**, 129–134.

Kreiser, A. M., Appleby, P. G., Natkanski, J., Rippy, B., and Battarbee, R. W. (1990). Afforestation and lake acidification: A comparison of four sites in Scotland. *Philosophical Transactions of the Royal Society, London* B **327**, 151–158.

Kreiser, A. M., and Battarbee, R. W. (1988). Analytical quality control (AQC) in diatom analysis. *Proceedings of Nordic Diatomist Meeting,* Report 12, Department of Quaternary Research, University of Stockholm, pp. 41–44.

Krishnaswamy, S., Lal, D., Martin, J. M., and Meybeck, M. (1971). Geochronology of lake sediments. *Earth and Planetary Science Letters* **11**, 407–414.

Kristiansen, J. (1986). Silica-scale bearing Chrysophytes as environmental indicators. *British Phycological Journal* **21**, 425–436.

Lawacz, W. (1969). The characteristics of sinking materials and the formation of bottom deposits in an eutrophic lake. *Mitteilungen der Internationalen Vereinigung für theoretische und angewandte Limnologie* **17**, 319–331.

Lehmann, J. T. (1975). Reconstructing the rate of accumulation of lake sediment: The effect of sediment focusing. *Quaternary Research* **5**, 541–550.

Liehu, A., Sandman, O., and Simola, H. (1986). Effects of peatbog ditching in lakes: problems in paleolimnological interpretation. *Hydrobiologia* **143**, 417–424.

Livingstone, D. (1984). The preservation of algal remains in recent lake sediments. *In* "Lake Sediments and Environmental History" (E. Y. Haworth and J. W. G. Lund, Eds.), pp. 191–202. Leicester University Press.

Longmore, M. E., O'Leary, B. M., and Rose, C. W. (1983). Caesium-137 profiles in the sediments of a partial meromictic lake on Great Sandy Island (Fraser Island), Queensland, Australia. *Hydrobiologia* **103**, 21–28.

Löffler, H. (1975). The onset of meromictic conditions in Goggausee, Carinthia. *Verhandlungen der Internationalen Vereinigung für theoretische und angewandte Limnologie* **19**, 2284–2289.

——. (1983). Changes in the benthic fauna of the profundal zone of Traunsee (Austria) due to salt mining activities. *Hydrobiologia* **103**, 135–140.

——. (1986a). Ostracod analysis. *In* "Handbook of Holocene Palaeoecology and Palaeohydrology" (B. E. Berglund, Ed.), pp. 693–702. Wiley, Chichester, England.

——. (1986b). An early meromictic stage in Lobsigensee (Switzerland) as evidenced by ostracods and *Chaoborus. Hydrobiologia* **143**, 309–314.

Lundquist, G. (1927). Bodenablagerungen and Entwicklungstypen der Seen. *Die Binnengewasser* **2**, 1–126.

Mackereth, F. J. H. (1969). A short core sampler for subaqueous deposits. *Limnology and Oceanography* **14**, 145–151.

Meriläinen, J. (1967). The diatom flora and the hydrogen-ion concentration of the water. *Annales Botanici Fennici* **4**, 51–58.

Moss, B. (1978). The ecological history of a medieval man-made lake, Hickling Broad, Norfolk. *Hydrobiologia* **60**, 23–32.

Munch, S. (1980). Fossil diatoms and scales of Chrysophyceae in the recent history of Hall Lake, Washington. *Freshwater Biology* **10**, 61–66.

Munro, M. A. R., Kreiser, A. M., Battarbee, R. W., Juggins, S., Stevenson, A. C., Anderson, D. S., Anderson, N. J., Berge, F., Birks, H. J. B., Davis, R. B., Flower, R. J., Fritz, S. C., Haworth, E. Y., Jones, V. J., Kingston, J. C., and Renberg, I. (1990). Diatom quality control and data handling. *Philosophical Transactions of the Royal Society, London* B **327**, 257–261.

Naumann, E. (1932). Grundzuge der regionalen Limnologie. *Die Binnengewasser* **11**, 176.

Nipkow, F. (1920). Vorlaufige Mitteilung über Untersuchungen des Schlammabsatzes im Zurichsee. *Schweiz Zeitschrift für Hydrobiologie* **1**, 100–122.

Nygaard, G. (1956). Ancient and recent flora of diatoms and Chrysophyceae in Lake Gribsø. Studies on the humic acid Lake Gribsø. *Folia Limnologica Scandinavica* **8**, 32–94.

Oksanen, J., Laara, E., Huttunen, P., and Meriläinen, J. (1988). Estimation of pH optima and tolerances of diatoms in lake sediments by the methods of weighted averaging, least squares and maximum likelihood, and their use for the prediction of lake acidity. *Journal of Paleolimnology* **1**, 39–49.

Oldfield, F. (1977). Lakes and their drainage basins as units of sediment-based ecological study. *Progress in Physical Geography* **1**, 460–504.

Oldfield, F., and Appleby, P. G. (1984). Empirical testing of ^{210}Pb-dating models for lake sediments. *In* "Lake Sediments and Environmental History" (E. Y. Haworth and J. W. G. Lund, Eds.), pp. 93–124. Leicester University Press.

O'Sullivan, P. E. (1983). Annually-laminated lake sediments and the study of Quaternary environmental changes—a review. *Quaternary Science Reviews* **1**, 245–313.

Pearsall, W. H. (1921). The development of vegetation in the English Lakes, considered in relation to the general evolution of glacial lakes and rock basins. *Proceedings of the Royal Society, London* B **92**, 259–284.

Pennington, W. (1943). Lake sediments: The bottom deposits of the North Basin of Windermere, with special reference to the diatom succession. *New Phytologist* **42**, 1–27.

Pennington, W., Cambray, R. S., and Fisher, E. M. (1973). Observations on lake sediments using fallout ^{137}Cs as a tracer. *Nature* **242**, 324–326.

Quennerstedt, N. (1955). Diatomeerna i Langans sjövegetation. *Acta Phytogeographica Suecica* **36**, 1–208.

Radle, N., Keister, C. M., and Battarbee, R. W. (1989). Diatom, pollen and geochemical evidence for the palaeosalinity of Medicine Lake, S. Dakota, during the late Wisconsin—early Holocene. *Journal of Paleolimnology* **2**, 159–172.

Renberg, I. (1976). Palaeolimnological investigations in Lake Prästsjön. *Early Norrland* **9**, 113–160.

——. (1981a). Improved methods for sampling, photographing and varve-counting of varved lake sediments. *Boreas* **10**, 255–258.

——. (1981b). Formation, structure and visual appearance of iron-rich, varved lake sediments. *Verhandlungen der Internationale Vereinigung für theoretische und angewandte Limnologie* **21**, 94–101.

——. (1986). Photographic demonstration of the annual nature of a varve type common in N. Swedish lake sediments. *Hydrobiologia* **140**, 93–95.

Renberg, I., and Hellberg, I. (1982). The pH history of lakes in southwestern Sweden, as calculated from the subfossil diatom flora of the sediments. *Ambio* **11**, 30–33.

Renberg, I., and Wik, M. (1984). Dating recent lake sediments by soot particle counting. *Verhandlungen der Internationale Vereinigung für theoretische und angewandte Limnologie* **22**, 712–718.

Renberg, I., Korsman, T., and Anderson, N. J. (1990). Spruce and surface-water acidification—an extended summary. *Philosophical Transactions of the Royal Society, London* B **327**, 371–372.

Renberg, I., and Wik, M. (1985). Carbonaceous particles in lake sediments—pollutants from fossil fuel combusion. *Ambio* **14**, 161–163.

Rodhe, W. (1969). Crystallization of eutrophication concepts in Northern Europe. *In* "Eutrophication: Causes, Consequences, Correctives," pp. 50–64. National Academy of Sciences, Washington, D.C.

Round, F. E. (1957). The late-glacial and post-glacial diatom succession in the Kentmere Valley deposit. I. Introduction, methods and flora. *New Phytologist* **56**, 98–126.

Saarnisto, M. (1979). Studies of annually laminated lake sediments. *In* "Palaeohydrological Changes in the Temperate Zone in the last 15,000 Years" (B. E. Berglund, Ed.), pp. 61–80. University of Lund, Sweden.

Schelske, C. L., Conley, D. J., Stoermer, E. F., Newberry, T. L., and Campbell, C. D. (1986). Biogenic silica and phosphorus accumulation in sediments as indices of eutrophication in the Laurentian Great Lakes. *Hydrobiologia* **143**, 79–86.

Shapiro, J. (1958). The core-freezer—a new sampler for lake sediments. *Ecology* **39**, 758.

Simola, H. (1977). Diatom succession in the formation of annually laminated sediment in Lovojärvi, southern Finland, during the past 600 years. *Annales Botanici Fennici* **14**, 143–148.

Smol, J. P. (1980). Fossil synuracean (Chrysophyceae) scales in lake sediments: A new group of paleoindicators. *Canadian Journal of Botany* **58**, 458–465.

Smol, J. P., Charles, D. F., and Whitehead, D. R. (1984). Mallomonadacean microfossils provide evidence of recent lake acidification. *Nature* **307**, 628–630.

Stevenson, A. C., Birks, H. J. B., Flower, R. J., and Battarbee, R. W. (1988). Diatom-based pH reconstruction of lake acidification using Canonical Correspondence Analysis. *Ambio* **18**, 228–233.

Stockner, J. G., and Benson, W. W. (1967). The succession of diatom assemblages in the recent sediments of Lake Washington. *Limnology and Oceanography* **12**, 513–532.

Swain, E. B. (1985). Measurement and interpretation of sedimentary pigments. *Freshwater Biology* **15**, 53–75.

ter Braak, C. J. F. (1986). Canonical correspondence analysis: A new eigenvector method for multivariate direct gradient analysis. *Ecology* **67**, 1167–1179.

———. (1987a). CANOCO—a FORTRAN program for canonical community ordination by partial detrended canonical correspondence analysis, principal components analysis and redundancy analysis (version 2.1). TNO Institute of Applied Computer Science, Wageningen.

———. (1987b). "Unimodal models to relate species to environment." Unpublished Ph.D. thesis, University of Wageningen.

ter Braak, C. J. F., and Barendregt, L. G. (1986). Weighted averaging of species indicator values: its efficiency in environmental calibration. *Mathematical Biosciences* **78**, 57–72.

ter Braak, C. J. F., and van Dam, H. (1989). Inferring pH from diatoms: A comparison of old and new calibration methods. *Hydrobiologia* **178**, 209–223.

Thienemann, A. (1925). Die Binnengewasser Mitteleuropas. Ein limnologische Einfuhrung. *Die Binnengewasser* **1**, 1–255.

Thompson, R. (1983). Global Holocene magnetostratigraphy. *Hydrobiologia* **103**, 45–52.

Tippett, R. (1964). An investigation into the nature of the layering of deep-water sediments in two eastern Ontario lakes. *Canadian Journal of Botany* **42**, 1693–1704.

Watts, W. A., and Winter, T. C. (1966). Plant macrofossils from Kirchner Marsh, Minnesota—a paleoecological study. *Bulletin of the Geological Society of America* 77, 1339–1359.

Whitehead, D. R., Charles, D. F., Jackson, S. J., Reed, S. E., and Sheehan, M. C. (1986). Late-glacial and Holocene acidity changes of Adirondack (N.Y.) lakes. *In* "Diatoms and Lake Acidity" (J. P. Smol, R. W. Battarbee, R. B. Davis, and J. Meriläinen, Eds.), pp. 251–274. W. Junk, Dordrecht.

Whiteside, M. C. (1983). The mythical concept of eutrophication. *Hydrobiologia* **103**, 107–112.

Whitmore, T. J. (1989). Florida diatom assemblages as indicators of trophic state and pH. *Limnology and Oceanography* **34**, 882–895.

Winter, T. C., and Wright, H. E., Jr. (1977). Paleohydrologic phenomena recorded by lake sediments. *Eos* **58**, 188–196.

Wright, H. E., Jr. (1967). A square-rod piston sampler for lake sediments. *Journal of Sedimentary Petrology* **37**, 975–976.

——. (1980). Cores of soft lake sediments. *Boreas* **9**, 107–114.

Züllig, H. (1961). Die Bestimmung von Myxoxanthophyll in Bohrprofilen zum Nachweis vergangener Blaualgenentfaltungen. *Verhandlungen der Internationalen Vereinigung für theoretische und angewandte Limnologie* **14**, 263–270.

——. (1982). Untersuchungen über die Stratigraphie von Carotinoidea im geschichteten Sediment von 10 Schweizer Seen zur Erkundung früherer Phytoplankton-Entfaltungen. *Schweizerische Zeitschrift für Hydrologie* **44**, 1–98.

6

Late Quaternary Climatic and Vegetational Change in Eastern North America: Concepts, Models, and Data

John E. Kutzbach
and
Thompson Webb III

Paleoclimatology

Paleoclimatic research is an interdisciplinary and international enterprise involving both data collection and modeling studies. The global and multivariate nature of the climate system requires the gathering and compilation of different types of data from all regions of the globe. The geological records about past climates are stored in many localities across the earth's surface from the bottom of the oceans to the tops of ice sheets and mountains, and the collection and correct interpretation of these data provide a challenge to field and laboratory researchers. Even when successful, paleoclimatologists face wide gaps in coverage and potentially contradictory information from different types of data. A general understanding of the climate system is therefore required to help in bridging the gaps and in resolving the contradictions. This understanding can be articulated as either qualitative concepts or formal mathematical models. The former are important guides to how, where, and what sorts of data should be collected and interpreted, whereas the latter are valuable tools both for generating new concepts and for translating theory into testable hypotheses about past climates.

The purpose of this chapter is to describe many of the concepts, new and old, that have guided and been generated by COHMAP (Cooperative Holocene Mapping Project, an interdisciplinary, international research group formed in 1974 by H. E. Wright, Jr., E. J. Cushing, J. E. Kutzbach, and T. Webb III). In a brief historical section, we note the ties between the global COHMAP perspective and that of European paleoclimatic researchers in the 1920s. Most of these concepts were derived from published research, data, analysis, and modeling studies and also from the interaction between model results and the data. The latter interaction is a recent development in COHMAP and represents a powerful method for checking published results, data interpretations, modeling results, and climate theory (COHMAP Members, 1988) (Fig. 1).

In this chapter, we will focus on the character of midcontinent Holocene climates to which our data and model results bring new understanding. We will describe several key concepts derived from our modeling studies and from taking a zoom-lens approach to displaying and interpreting pollen data. Finally, we will present comparisons of different types of data and of data and model results.

COHMAP Research

Prolonged and friendly interaction with critical and knowledgeable colleagues is an essential ingredient to successful interdisciplinary research. Meetings among COHMAP researchers have provided a forum for such interactions. Paleoclimatic research proceeds best as a community effort in which researchers of different specialties can test new ideas under the guidance of those with critical understanding of the different types of data, theories, or models. Like CLIMAP (Climate: Mapping, Analysis and Prediction) (1976) before it, COHMAP has been one attempt to provide such a research environment, but we have also profited from many interactions with researchers outside of COHMAP.

COHMAP set as its goal to advance our knowledge of climatic change since the last glacial maximum (ca. 18,000 yr B.P.) by answering these broad questions: What happened? When did it happen? Why did it happen? The answers to *what* and *when* came from fieldwork and literature research. Ideas about *why*

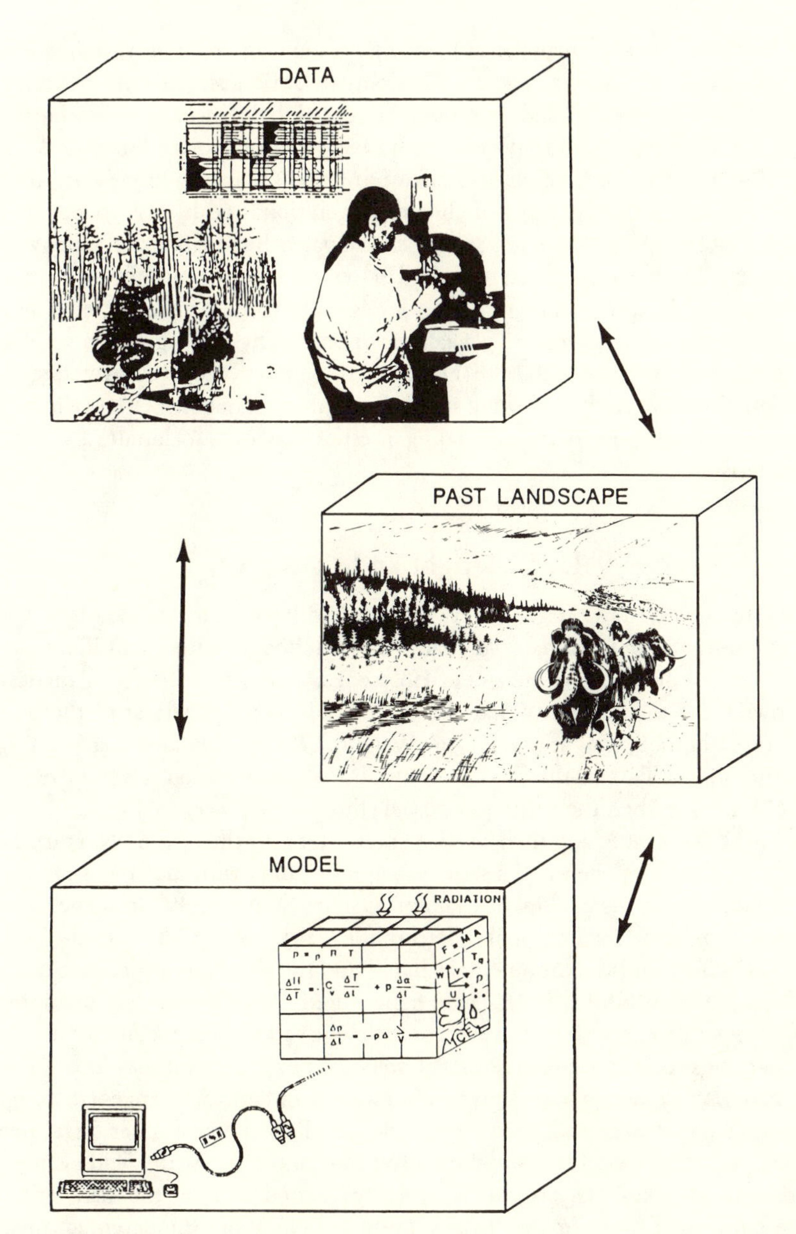

Fig. 1. A schematic view of elements of COHMAP that incorporates, in the past landscape box, the symposium picture that was the integrative, interdisciplinary theme of the tribute to the Limnological Research Center and Herb Wright. The other elements of COHMAP, less poetic or artistic, perhaps, are detailed analyses of geologic data and numerical simulations of paleoclimates using supercomputers.

came from generating hypotheses based on geologic observations and then testing the hypotheses with numerical climate models. Because patterns of climate, such as planetary waves and monsoons, have a global structure, we decided to take a global view. We also decided to study a series of climatic "snapshots," at 3,000-year intervals, from glacial maximum 18,000 years ago to present, in order to illustrate the dynamics of climate, vegetation, and hydrologic changes. COHMAP has helped to clarify several classic paleoclimatic puzzles. Why were lake levels low in the midwestern United States but high in parts of North Africa in the mid-Holocene? Why were lake levels high in the American Southwest, and low in the Northwest, at glacial maximum? Why did vegetation with no modern analogs grow 18,000 to 12,000 yr B.P. and why did taxa migrate individualistically? In the following sections we will explain how COHMAP helped clarify or answer some of these puzzling questions about past climates and vegetation.

H. E. Wright and COHMAP

Long before COHMAP began, H. E. Wright had been asking what, when, and why questions about climate. He had helped define, describe, and illuminate many of the outstanding puzzles of Holocene climate. In the North American Midwest, he and coworkers had charted the advance and retreat of the prairie/forest border (Wright *et al.*, 1963; Wright, 1968; Wright and Watts, 1969). In the Middle East in the Zagros Mountains of Western Iran (Wright *et al.*, 1967), he described the entire postglacial climatic sequence: cool and dry until about 11,000 yr B.P. and then warmer and wetter over the next 6,000 years. Finally, in the 1970s, he wrote a series of papers dealing with the large-scale patterns and the timing of climatic change: "Dynamical Nature of Holocene Vegetation" (1976a); "Environmental Setting for Plant Domestication in the Near East" (1976b); and "Environmental Change and the Origin of Agriculture in the Old and New World" (1977). These papers used paleoclimatic data to address questions of climate change, vegetation change, and human adaptations.

Wright's contributions to COHMAP went far beyond his publications. In informal discussions, he shared his vast knowledge of climatic changes stemming from his own travels and from his knowledge of the work of other researchers in all parts of the world. His ability to combine and contrast the field evidence from pollen, lakes, glaciers, dunes, and bogs helped us to see the many effects and patterns of past climate changes. His questions about the potential consequences of this or that change in the height of the ice sheet, the position of the jet stream, or the shift of the sea-ice border stretched our climatic thinking. His love and enthusiasm for Quaternary climates—and for those with like interests—helped stimulate much COHMAP research.

Historical Review

ORBITAL CHANGES: PACEMAKERS OF ICE AGES AND MONSOONS

Today we recognize how orbital variations have influenced both high-latitude and tropical climates. This recognition took time to develop. The idea that the orbital changes of tilt, precession, and eccentricity helped to fix the timing of ice ages is rooted and developed in the work of Adhémar (1842), Croll (1864), and Milankovitch (1920). Imbrie and Imbrie (1979) have traced the evolution of this idea from the time of its origin to the present. A crucial step was the demonstration by Hays *et al.* (1976) that the periodicity of climatic oscillations (indicated by geological data with stratigraphic and radiometric dating) agreed with the periodicities of orbitally induced changes in insolation. Since then a general concensus has developed that glacial-interglacial cycles are paced by the insolation variations induced by the orbital cycles. What has remained unclear, however, is the variety of mechanisms linking cause and effect. Milankovitch (1920) among others proposed that cool summers in high latitudes would favor the preservation of the previous winter's snowfields and that these would eventually form ice sheets. Detailed testing and quantitative confirmation of this and other mechanisms is still an elusive goal, in part because climate is a multivariate global system with many interactions and feedbacks.

Changes in tropical climates have also been linked to orbital variations. Spitaler (1921) proposed that orbital changes would influence tropical climate by altering the temperature contrast between land and sea and thereby influencing the strength of monsoons. Various researchers found evidence of orbital periodicities in marine sediments (Rossignol-Strick, 1983; Prell, 1984a,b; Pokras and Mix, 1985), and research within COHMAP has advanced our understanding of the detailed mechanisms linking orbital change and tropical monsoon changes (Kutzbach, 1981; see below).

The study of the mechanisms of climatic change is complicated by the climate system's many components (air, ice, water, land, biosphere) and the wide range of characteristic timescales for various interactions among these components. Huge ice sheets, for example, take many millennia either to build up or to melt. Imbrie and Imbrie (1980) estimate that time constants of about 42,000 years (growth phase) and 10,600 years (decay phase) fit the geologic record of inferred changes of ice volume during the past 700,000 years. Understanding of the climate of a particular time, such as the early Holocene, therefore requires researchers to consider not only the possible "external" factors that might be influencing the climate directly (such as the current insolation regime), but also the residual influences of earlier events "transmitted" by the slowly responding elements of the climate system itself (e.g., melting of ice sheets, changes of sea level, crustal rebound, etc.—see Saltzman, 1985). The following section reviews how Quaternary scientists have struggled with these notions of causes, mechanisms, and internal feedbacks in their quest to explain the evolution of climate since the last major glaciation almost 18,000 years ago.

Brief Review of Paleoclimatic Research Since the 1920s

Historical Background

The 1920s were marked by exciting interactions among climatologists, geologists, and mathematical modelers of climate. Their attempts to link data with model results incorporating astronomical forcing were similar in spirit to recent interactions fostered by COHMAP and SPECMAP (Spectral Mapping Project) (Kerr, 1984, 1986). The work during the 1920s was influenced by the early pioneering work on orbital theories and ice ages done by Croll and Milankovitch (see Imbrie and Imbrie, 1979).

In 1920, Milankovitch published his first monumental and quantitative treatment of the seasonal and latitudinal changes of solar radiation associated with orbital changes. The book, written in French, also contained his estimates of the climatic (temperature) response to insolation change (Milankovitch, 1920). Milankovitch obtained these estimates by developing and using a zonal-average energy-budget model. His was a pioneering effort, but one particularly significant weakness of the model was that it did not allow for lateral heat transport (see Kutzbach, 1985). He was the first to apply a numerical climate model to estimating the climatic impact of orbital forcing. He found that summers could have been 5 °C warmer and winters 5 °C colder with perihelion in northern summer and with increased axial tilt (at 24° versus 23 1/2° today, i.e., conditions applicable to about 10,000 years ago). Spitaler (1921) worked independently of Milankovitch and obtained similar results by using a more empirical climate model than that of Milankovitch. For example, in Spitaler's model, long-wave (infrared) radiation was estimated empirically rather than with the physical radiation equations.

Almost immediately, a fruitful exchange of ideas occurred among Milankovitch, Köppen (one of the best-known meteorologists/climatologists of the time), and Wegener (a meteorologist/geophysicist and Köppen's son-in-law). Köppen and Wegener's 1924 book, *Die Klimate der geologischen Vorzeit,* strongly supported the link between orbital cycles and glacial cycles; this support, coming from established and respected climatologists, helped establish the importance of the "Milankovitch theory" of ice ages.

Köppen and Wegener (1924) also realized that the climatic record includes not only the direct effects of orbital changes but also delayed effects because of the existence of ice sheets with their long time-constants or response-times (Fig. 2). They had used the observational record of vegetation change in Europe (Fig. 2) to infer the temperature record of the past 25,000 years. Their concept

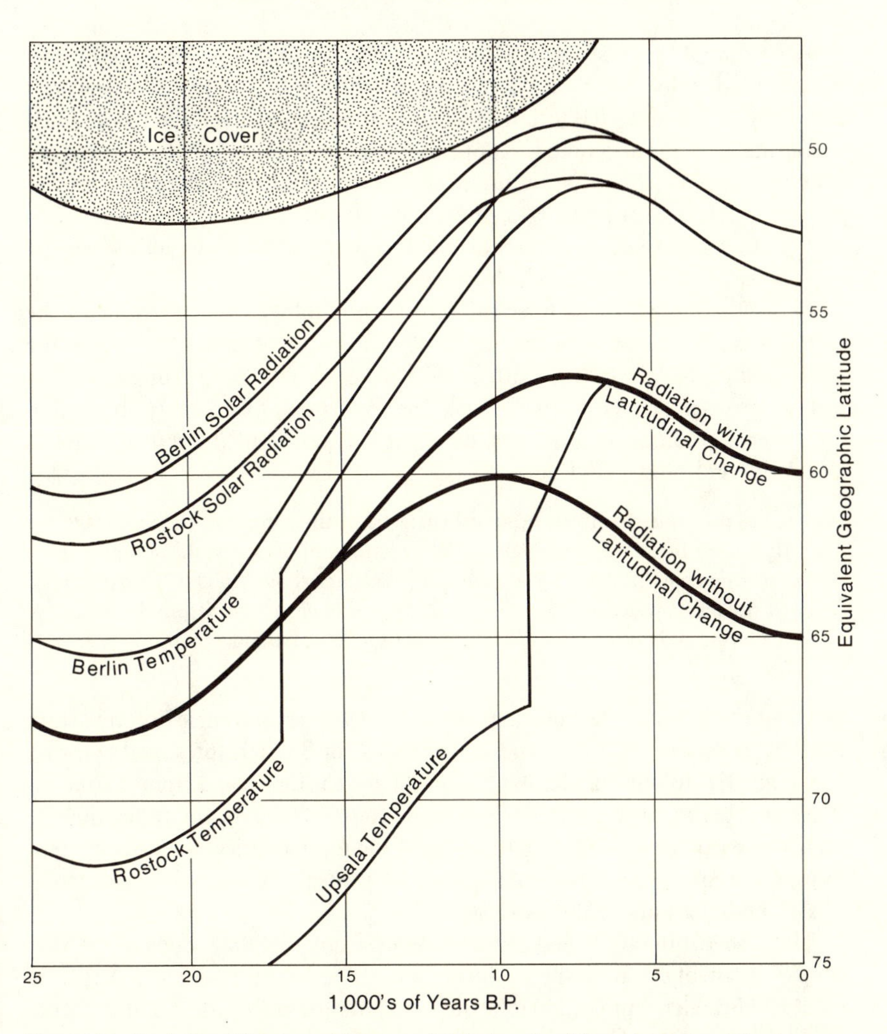

Fig. 2. This sketch from Köppen and Wegener (1924) (translated from the German) incorporates conceptually the effects of ice sheets, orbitally induced changes of solar radiation for Berlin and Rostock, and the climatic response (Berlin, Rostok, and Upsala temperatures). The authors also considered polar wandering effects ("Radiation with latitudinal change" and labeled "Stralung, mit Breitenabnahme" on the original) now thought to be very small. The radiation forcing and temperature effects are scaled to an "equivalent geographic latitude." The timescale is thousands of years before present.

was that the cold periglacial climate from 25,000 to 10,000 years ago combined with the radiatively induced, warmer summers from 10,000 to 5,000 years ago could explain the inferred record of temperature from geological data obtained in Germany. They also noted (Köppen and Wegener, 1924: 250) that the so-called postglacial climatic optimum would have had both warmer summers and colder winters than at present—a direct consequence of the orbital changes (see Sec. 3), but a consequence often overlooked in subsequent decades when the popularity of the Milankovitch theory waned.

C. E. P. Brooks, the influential British climatologist also working in the 1920s, was likewise aware of the importance of the astronomical theory and the delayed effects of glacial-age features (such as ice sheets) for explaining the record of postglacial climates. In his book *The Evolution of Climate* (published in 1922, second edition in 1925), he describes the period of the retreat of ice as a "continental phase" and notes,

> It seems probable that the continental character of the climate of the final stages of the retreat phase was slightly increased by astronomical causes, the obliquity of the ecliptic being probably nearly one degree greater about 7,500 B.C. than it is now. In Germany and Sweden this would have the effect of lowering the winter temperature and raising the summer temperature by rather more than 1°F. [1922: 120]

He notes that this effect of colder winter and warmer summers (inferred from fossil evidence) began earlier in Germany than in Sweden, presumably owing to the greater distance of Germany from the retreating ice and from marine influences. He described other likely climatic consequences of the retreating ice, such as the expected decreasing intensity of the glacial anticyclone over western Europe (Brooks, 1922: 122). According to Brooks (1922), this "continental phase" ended around 6,000 years ago.

The above quotation and other sections from his 1922 book show that Brooks's grasp of the astronomical theory was still rather limited. He quotes from Croll and Spitaler, but not Milankovitch, and mentions the astronomical theory only in the context of glacial cycles. But by 1926 when he published his major book, *Climate through the Ages,* he had read Milankovitch (1920), Köppen and Wegener (1924), and other studies; and he produced a comprehensive and physically sound view of the causes of and evidence for evolution of late Quaternary climates, including Spitaler's ideas on orbital cycles and monsoon (pluvial) cycles in the tropics.

From the 1920s to the 1960s, further progress of this sort slowed, and the popularity and application of the astronomical theory diminished. Several factors were probably important: (1) In 1940, Sir George Simpson, then of the Royal Meteorological Society, published a paper that concluded, "the changes of solar radiation due to changes in the Earth's orbit are always too small to be of practical importance" (Simpson, 1940: 209; see Kutzbach, 1985, for interpretation of how Simpson reached this incorrect but influential conclusion). (2) Before widespread application of radiometric dating, no independent check was possible for the timescale of the fossil evidence at most sites (independent of the solar radiation curves), the exception being those few sites with varve (annual-lamination) chronologies. (3) The development of quantitative climate models (beyond Milankovitch's early work) awaited the invention of high-speed computers that allowed development of general circulation models and improvements of energy-budget models. These developments began to have an impact in the 1960s when, for the first time, quantitative reasoning about the mechanisms of climate would advance beyond the stages reached in the 1920s.

Some of the key new approaches that began to emerge by the 1960s and 1970s include the following.

Marine Sediment Cores

The astronomical theory predicts that if the climate system (ice-oceans-atmosphere) responds quasi-linearly to external forcing, then geologic records should contain cycles with periods of around 22,000 years (precession), 41,000 years (tilt), and 100,000 years (eccentricity). Long cores of marine sediment dated magnetostratigraphically, first retrieved in the 1950s, began to support the astronomical theory (e.g., Emiliani, 1955; Broecker *et al.*, 1968), as did reef terraces dated radiometrically (Mesoella *et al.*, 1969). Also crucial was the understanding that $\delta\ ^{18}O$ variations in the foraminifera from the deep sea were primarily a record of global ice volume (Shackleton and Opdyke, 1973). Following a decade of growing evidence in support of the astronomical theory, Hays *et al.* (1976) published detailed time analyses confirming that cycles of about 22,000, 41,000, and 100,000 years were widespread in marine (isotopic) records. They concluded that orbital changes indeed are "pacemakers" of glacial/interglacial periods during the ice ages.

General Circulation Theory

The development of general circulation theory, concepts, and models in the 1950s and 1960s made possible a wide range of quantitative climatic studies (see Kutzbach, 1985). Quantitative solutions emerged to problems of airflow over and around mountains and plateaus, monsoonal circulations, and the three-dimensional structure of atmospheric motions (Washington and Parkinson, 1986). By the mid-1960s, climatologists were poised to undertake quantitative studies of climatic change. This perspective is clearly expressed in J. M. Mitchell's chapter on "Theoretical Paleoclimatology" (Mitchell, 1965). These new perspectives on the laws of atmospheric motions and thermodynamics also aided Lamb *et al.* (1966), Lamb and Woodroffe (1970), and Bryson and Wendland (1967) in estimating the likely flow patterns in the vicinity of ice sheets and in comparing mapped geologic evidence with climatic patterns inferred from climate theory.

CLIMAP

CLIMAP (Climate: Mapping, Analysis and Prediction) was the first research group to organize a global paleoclimatic experiment aimed at quantitative comparison of well-dated and climatically calibrated geologic evidence with paleoclimate simulations by general circulation models. The development of transfer-function methods for the climatic calibration of marine-plankton data was key to CLIMAP research (Imbrie and Kipp, 1971). The multiauthored publications (CLIMAP, 1976, 1981) described their reconstructions for the state of the oceans, land surface, and ice sheets at full-glacial time (about 18,000 yr B.P.). CLIMAP members' collaborative efforts with climate modelers began to answer questions about the climatic mechanisms that were operating during the last glacial maximum (Gates, 1976; Manabe and Hahn, 1977). The questions included: (1) What was the circulation in the vicinity of the ice sheets? (2) Where were the storm tracks? (3) How did tropical precipitation patterns at 18,000 yr B.P. differ from those today? (4) How intense was the Asian monsoon in summer?

COHMAP

COHMAP was designed as an interdisciplinary research project that would describe the evolution of Holocene climates and explore the dynamical and thermodynamical mechanisms that produced those changes. It took form first in 1977 as an NSF-sponsored COHMAP grant with E. J. Cushing, J. E. Kutzbach, A. M. Swain, T. Webb III, and H. E. Wright, Jr., as co-Principal Investigators. The initial idea was to concentrate on eastern North America and on the period from 12,000 yr B.P. to present with the prairie/forest border as a key paleoclimatic indicator. The project, however, gradually expanded to aim at global data sets (Peterson *et al.*, 1979; Webb, 1985) and at understanding the mechanisms of the general evolution of climate from 18,000 yr B.P. to present (Kutzbach, 1981; COHMAP Members, 1988). With this latter development, P. J. Bartlein, F. A. Street-Perrott, W. L. Prell, and W. F. Ruddiman became major contributors to COHMAP research along with a large group of palynologists and Quaternary scientists from the U.S. and abroad.

Small, informal meetings in St. Paul, Madison, and Providence helped to shape the key ideas and tasks that guided COHMAP research. These included (1) the compilation of global sets of well-dated paleoclimatic data that provided synoptic views of past climates (Fig. 3); (2) the development of hydrologic budget models and multivariate statistical methods for the calibration of the data in climatic terms (Kutzbach, 1980; Howe and Webb, 1983; Bartlein *et al.*, 1986); (3) the use of the "COHMAP diagram" (Fig. 4), a schematic reminder that (a) solar radiation changes caused by orbital variations and (b) long-lasting

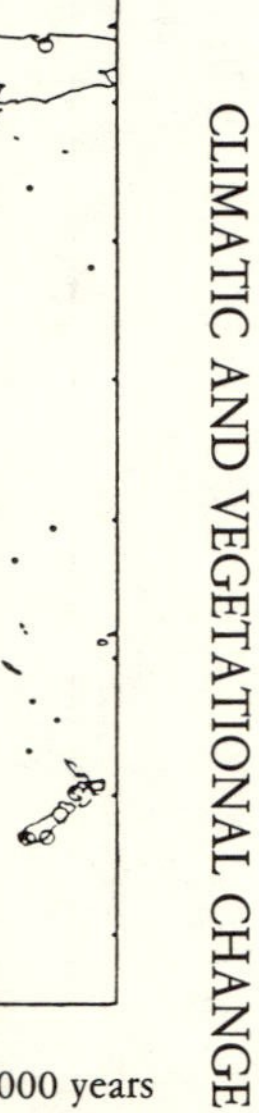

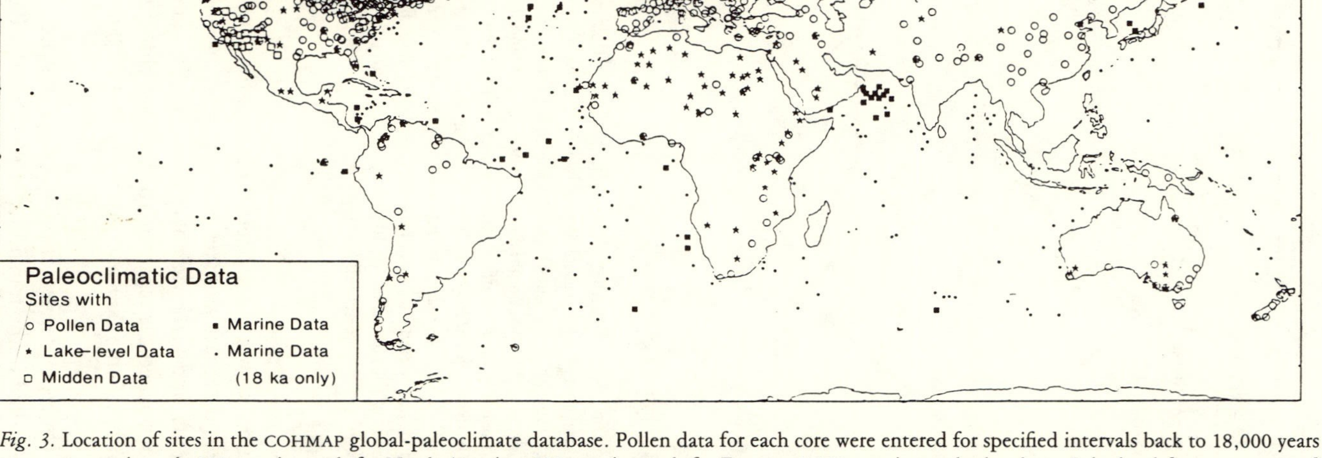

Fig. 3. Location of sites in the COHMAP global-paleoclimate database. Pollen data for each core were entered for specified intervals back to 18,000 years ago: at approximately 300-year intervals for North America, 500-year intervals for Europe, 3,000-year intervals elsewhere. Lake-level data were entered for each 1,000-year interval, and marine plankton for each 3,000-year interval. Radiocarbon-dated, plant-macrofossil assemblages from packrat middens were entered in 3,000-year intervals. The location of CLIMAP marine-plankton records for 18 ka are also indicated. (With permission: the AAAS; copyright 1988 the AAAS.)

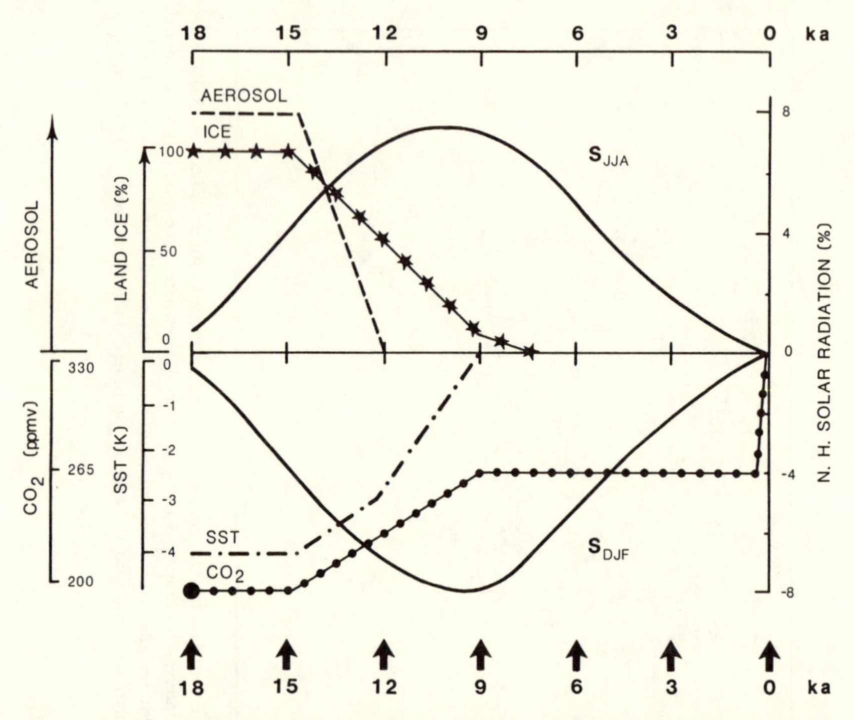

Fig. 4. Boundary conditions for the COHMAP simulation with the Community Climate Model for the last 18 ka (thousand years ago). External forcing is shown for Northern-Hemisphere solar radiation in June through August (S_{JJA}) and December through February (S_{DJF}) as percent difference for the radiation at present. Internal boundary conditions include land ice (Ice) as percent of 18 ka ice volume; global, mean annual sea-surface temperatures (SST) as departures from present; excess glacial-age aerosol (Aerosol) with arbitrary scale; and atmospheric CO_2 concentration (CO_2) in parts per million by volume. The arrows mark the times of the seven sets of simulation experiments. The one for 18 ka included the lowered CO_2 concentration (200 ppmv, large filled circle); the others had the CO_2 concentrations of the control case (330 ppmv) rather than a stepwise increase. Experiments incorporating increased glacial-age aerosol loading have not been completed. (Reprinted with permission: *Nature* 317: 130.)

glacial-age features such as slowly retreating ice sheets were important determinants of late Quaternary climates and could be incorporated into climate-model experiments, and 4) the view of climate, observed and simulated, in time and space that permitted the representation of past climates at a global scale in a series of "snapshots" in 3,000-year intervals from 18,000 years ago to present (Fig. 5).

Certain key concepts derived from COHMAP research involving modeling and data are described in the remainder of this chapter. Our discussion focuses on a subset of the COHMAP results—namely, our improved understanding of the evolution of climate and vegetation in the early to mid-Holocene in central North America. To set the stage for this focus, we briefly review previous studies on the period often called the hypsithermal (occurring during early to mid-Holocene time) in North America.

Hypsithermal

The mid-Holocene interval from 8,000 to 4,000 years ago is often referred to as the altithermal or hypsithermal, and these terms imply that the global as well as regional mean temperatures were higher then than at present. Many sites provide evidence that local and regional temperatures were higher than today (Davis *et al.*, 1980; Bartlein and Webb, 1985; Huntley and Prentice, 1988), but no clear evidence exists that the global mean temperature was higher than at present (Webb and Wigley, 1985). Deevey and Flint (1957) formally defined the term *hypsithermal* at a time when Quaternary scientists believed that evidence from local sites could indicate when the period of maximum global warmth had occurred during the present interglacial period. Terms like *altithermal* (Antevs, 1947), *xerothermic* (Sears, 1942), and *climatic optimum* have also been used to describe this climatic episode. Deevey and Flint (1957) defined the climatostratigraphic term *hypsithermal* (period of high temperatures) as a chronostratigraphic unit synchronous with the formation of certain Danish pollen (i.e., biostratigraphic) zones. This definition works only if the biostratigraphic zones indicating maximum temperatures in Denmark occur at a time when temperatures were higher than at present in enough other regions of the world that the global mean temperature was higher than at present.

Since 1957, the chronostratigraphic usage (i.e., reference to a specific time period) of *hypsithermal* has been neither strict nor frequent, and its most common usage has been as a regional climatostratigraphic term (i.e., in reference to a climatic condition) of somewhat vague definition (Webb and Wigley, 1985). Some researchers interpret this period as being hotter or drier or both than today (Wright, 1976a). In fact, Watson and Wright (1980) have demonstrated why a term like *hypsithermal* cannot be applied both chronostratigraphically and climatostratigraphically when descriptive of local and regional conditions. Clearly, we are at a stage in paleoclimatic research when the term *hypsithermal* either requires a new definition or should be abandoned.

Concepts from COHMAP Model Studies

Our historical review showed that many ideas existed about the ultimate climatic consequences of orbitally induced changes in insolation and about the climatic effects of ice sheets on nearby regions. As early as 1920, Milankovitch had employed a simplified mathematical model of earth's climate based on the laws of

Present
6 ka
9 ka
12 ka
subtropical high (July)
westerly surface winds
January jet stream
southerly surface winds (July)
Present
stronger subtropical high (July)
warmer, drier than present (July)
stronger flow (July)
stronger flow (Jan)
warmer, drier than present (July)
stronger flow (July)
6 ka
stronger subtropical high (July)
warmer, drier than present (July)
warmer, drier than present (July)
stronger flow (Jan)
warmer, drier than present (July)
glacial anti - cyclone (July)
cooler, moister than present (July)
stronger onshore flow (July)
warmer than present (July)
9 ka
weaker subtropical high (July)
colder than present
stronger flow (July)
stronger flow (Jan)
glacial anticyclone (July)
colder, drier than present
colder than present
very strong westerly flow
12 ka

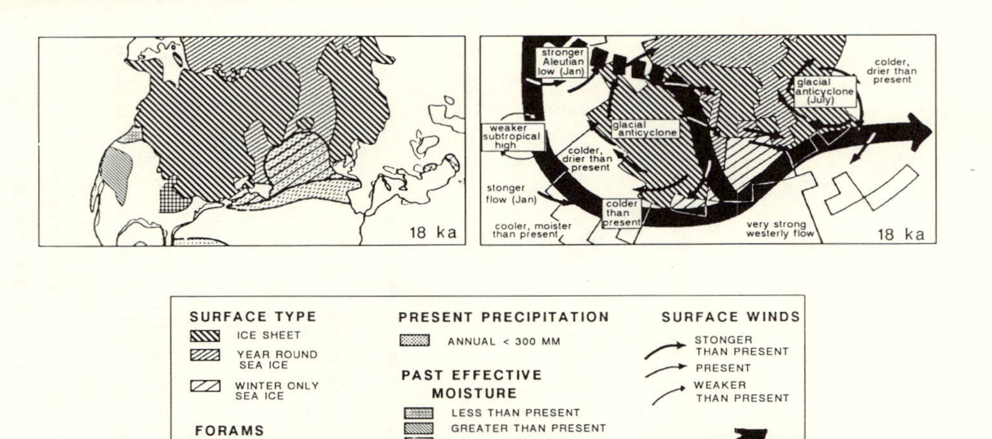

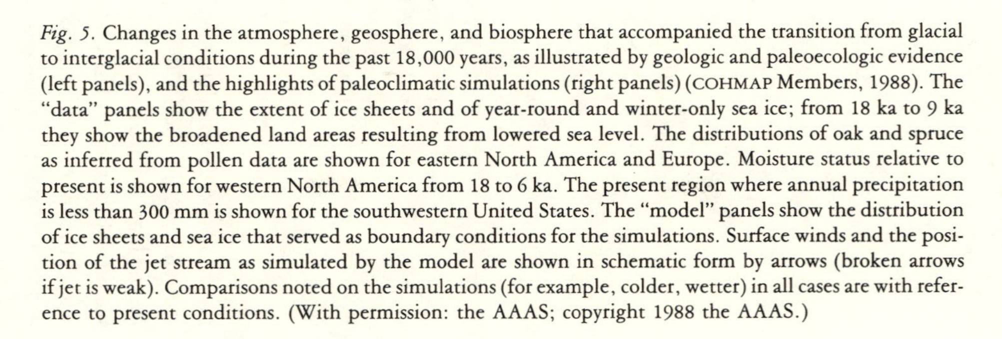

Fig. 5. Changes in the atmosphere, geosphere, and biosphere that accompanied the transition from glacial to interglacial conditions during the past 18,000 years, as illustrated by geologic and paleoecologic evidence (left panels), and the highlights of paleoclimatic simulations (right panels) (COHMAP Members, 1988). The "data" panels show the extent of ice sheets and of year-round and winter-only sea ice; from 18 ka to 9 ka they show the broadened land areas resulting from lowered sea level. The distributions of oak and spruce as inferred from pollen data are shown for eastern North America and Europe. Moisture status relative to present is shown for western North America from 18 to 6 ka. The present region where annual precipitation is less than 300 mm is shown for the southwestern United States. The "model" panels show the distribution of ice sheets and sea ice that served as boundary conditions for the simulations. Surface winds and the position of the jet stream as simulated by the model are shown in schematic form by arrows (broken arrows if jet is weak). Comparisons noted on the simulations (for example, colder, wetter) in all cases are with reference to present conditions. (With permission: the AAAS; copyright 1988 the AAAS.)

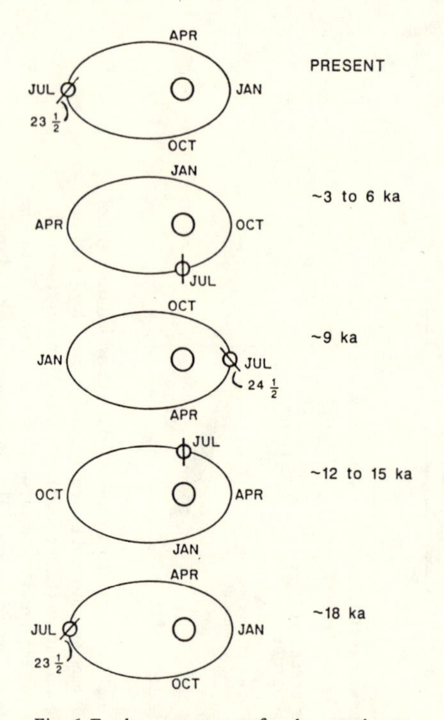

Fig. 6. Earth-sun geometry for about 18 ka, 15 to 12 ka, 9 ka, 6 to 3 ka, and present. Note that perihelion is in January at present and in July around 9 ka. Also note the increased tilt (24 ½ °) at 9 ka compared to the 23 ½ ° tilt at 18 ka and present.

radiation and thermodynamics to translate the insolation changes into temperature changes. By the 1970s, general circulation models of the atmosphere based on the laws of fluid motion were being used to simulate the flow patterns in the vicinity of ice sheets. The COHMAP project built on these previous studies but added the key concept (employed independently much earlier by Spitaler, 1921) of the differential response of land and ocean to orbitally produced changes in insolation. This COHMAP research helped bring the concepts of monsoon circulations, continentality, and seasonality to the foreground as part of the Milankovitch puzzle and stimulated interest in understanding the changes in both tropical and mid-latitude climates when describing the changes from glacial to interglacial conditions (Kutzbach and Street-Perrott, 1985; Prell and Kutzbach, 1987).

Changes in earth-sun geometry are at the core of our understanding of late Quaternary climate change. These changes are expressed compactly in the form of equations, tables, and time-series graphs (Milankovitch, 1920; Berger, 1978), but the changing earth-sun geometry is best illustrated through a series of pictures (Fig. 6) showing several orbital configurations for 18,000, 15,000–12,000, 9,000, and 6,000–3,000 years ago and present. The earth-sun geometry of 18,000 years ago was similar to that of the present. Between about glacial-maximum time and 9,000 yr B.P., the time of perihelion changed from January to July, advancing about one day every 60 years, and the axial tilt increased. Since about 9,000 yr B.P., axial tilt decreased, and the time of perihelion advanced through the remainder of the year, until today it occurs during the northern winter.

These orbital changes combined to produce the changes in the seasonal cycle of solar radiation portrayed in Fig. 7, here summarized in terms of Northern-Hemisphere averages (see also Fig. 4). Compared to our present regime, the seasonality of solar radiation was considerably increased from around 12,000 to 6,000 yr B.P. (Fig. 7, top), with maximum increase around 9,000 yr B.P. when there was more solar radiation in northern summer. Insolation was about $30W/m^2$ or about 7% more than it is today, and winter insolation was less (Fig. 7, middle). In the Southern Hemisphere the seasonality of solar radiation was decreased around 9,000 yr B.P. A key concept is that the small heat capacity of land (North America, North Africa/Eurasia), compared to ocean (the North Pacific and Atlantic), causes the response of surface temperature to the insolation changes to be much larger over land than over ocean (Fig. 7, bottom). Realistic values of heat capacity for land and for a 100-m mixed-layer ocean (that part of the ocean that experiences the largest seasonal cycle) indicate that northern continental interiors might have been as much as 5°C warmer than at present in summer around 9,000 yr B.P. (and generally 2–4°C warmer from 12,000 to 6,000 yr B.P.), whereas the northern oceans were warmer by about 1°C. The seasonal warming of the ocean would also lag behind the warming of the land (Fig. 7, bottom). Northern winters would be correspondingly colder, resulting in increased seasonality and continentality of climate (with the reverse holding for southern continents), but with perhaps little if any change in annual average temperature. The magnitude of these temperature changes was confirmed in our series of modeling experiments (Kutzbach and Otto-Bliesner, 1982; Kutzbach and Guetter, 1986; Kutzbach and Gallimore, 1988).

Based on hydrostatic considerations, the relatively large increase of summer temperature over northern land, compared to ocean, leads to higher air temperatures over land, lower surface pressure (compared to over ocean), and inflow of air from ocean to land at low levels (Fig. 8)—that is, an increased summer monsoon. The increased inflow causes increased precipitation and moister con-

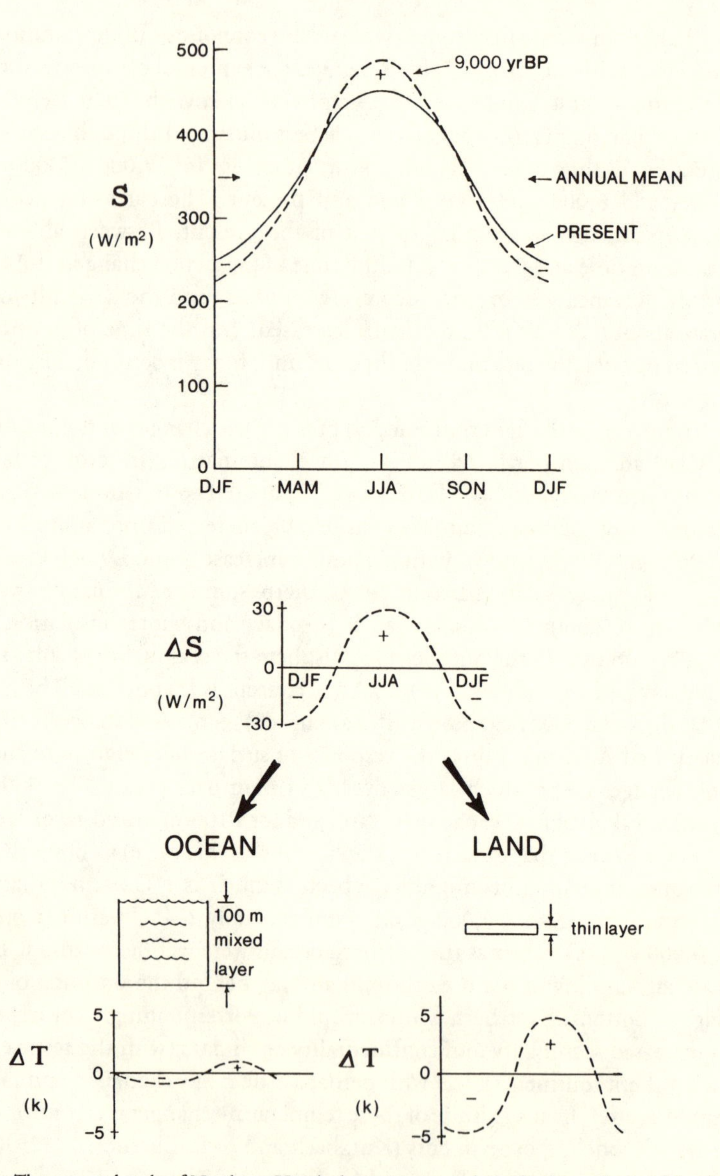

Fig. 7. Top: The seasonal cycle of Northern-Hemisphere, average solar radiation, calculated for the top of the earth's atmosphere, in W/m^2. Solid line (present); dashed line (9,000 yr B.P.). Middle: The changed input of solar radiation at 9,000 yr B.P. compared to present. Bottom: The schematic temperature response of a 100-m slab of ocean (left) and the insulative land surface (right) to the same change in solar radiation. It is this differential thermal response of land and ocean to changes of the seasonal cycle of solar radiation that drives the global monsoonal response to changes in earth's orbit.

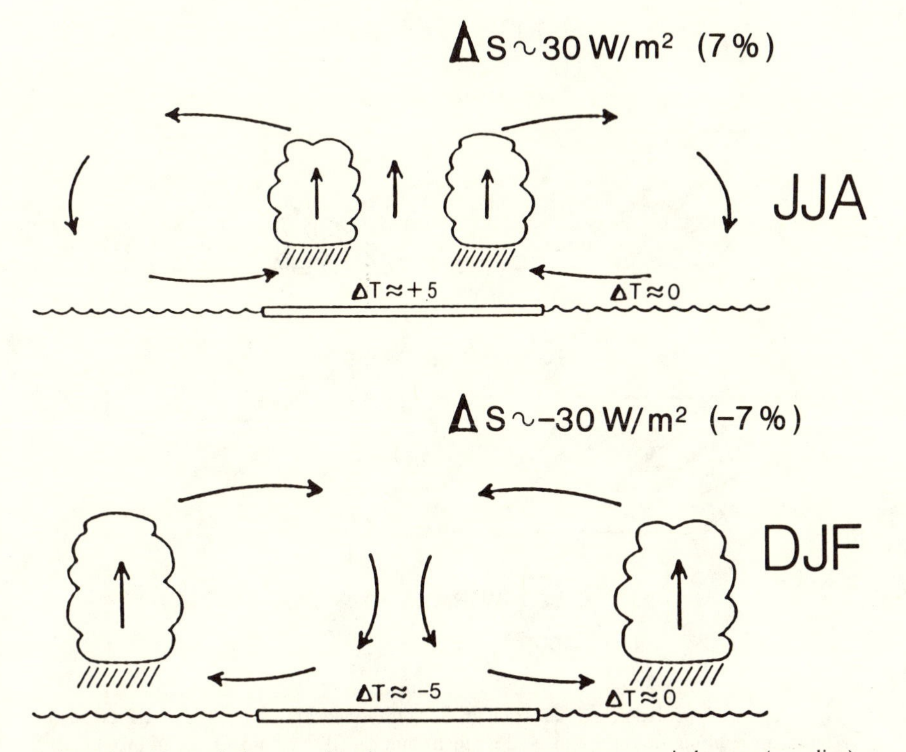

Fig. 8. Schematic, vertical cross section along an east-west transect across tropical oceans (wavy line) and land (thin slab). The increased temperature over land in northern summer (JJA) at 9 ka causes increased rising motion, low-level inflow of moist air, and high level outflow. Enhanced monsoon precipitation occurs along the coasts and to a certain distance inland. The interior is hotter but does not benefit from increased precipitation and therefore becomes drier. In northern winter (DJF) the reversed vertical-circulation cell develops. These *changes* (at 9 ka) enhance the *normal* vertical (monsoonal) cells.

ditions along the coasts and some distance inland; but far inland, beyond the reach of the increased flux of moisture, the climatic conditions may be drier owing to increased evaporative losses (Fig. 9; Kutzbach and Otto-Bliesner, 1982; Kutzbach and Guetter, 1986; Kutzbach and Gallimore, 1988). Our modeling studies with interactive soil moisture (Gallimore and Kutzbach, 1989) showed that these differential effects of moistening and drying are more pronounced when soil-moisture feedback is included.

By analogous reasoning, monsoon circulations for northern winters would also be intensified (Fig. 8); however, net annual moisture (precipitation-evaporation, P-E) tends to increase because the increase of summer moisture outweighs any decrease of winter moisture. This increase is yet another impor-

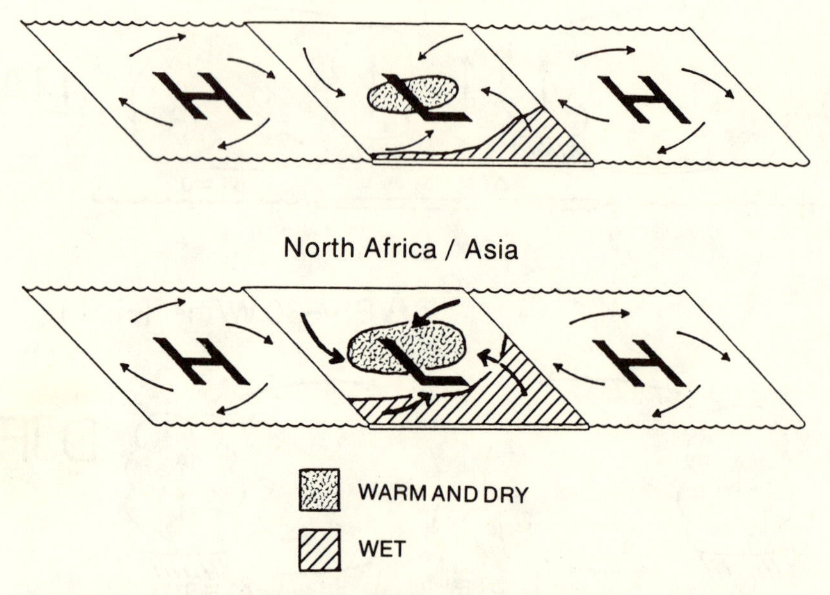

Fig. 9. Idealized circulation and climate on small (North America) and large (North Africa/Asia) continents resulting from the enhanced northern summer (JJA) monsoons caused by orbitally induced changes in solar radiation. The temperature in the interior should increase most over the large continent (as shown by the size of the region of greatest warmth) and the region of increased moisture should extend farther into the interior on the south and east sides of the large continent (as shown by the size of the region of greatest precipitation). The monsoon circulation is also more intense on the large continent (as shown by the thickness of wind arrows).

tant concept and results, in part, because of the nonlinearity between temperature and vapor pressure, which is represented by the Clausius-Claperon equation.

If solar radiation changes were the only causal factor to be considered in Holocene climatic change, one would arrive at a simple conceptual model where the magnitude of the "monsoonal" change depends only on the size and perhaps latitude of the continent. Size is a factor because the maximum temperature response mentioned earlier (about 5 °C) occurs only if the continent is so large that the interior is isolated from the moderating (advective) influence of the cooler ocean. The importance of the size of continents for determining the seasonal extremes of temperature is clearly demonstrated in experiments with energy-budget climate models by Crowley *et al.* (1986). Thus, an "idealized" view of northern monsoon enhancement at 9,000 yr B.P. would anticipate a larger temperature, moisture, and circulation response for North Africa/Eurasia than for North America (Fig. 9).

The sketch of summertime circulation and climate at 9,000 yr. B.P. over North Africa/Eurasia (Fig. 9, bottom) captures the essence of the observed climate around 12,000–6,000 yr B.P. in the large North-African/Eurasia land mass. The observations of moister climate at this time are primarily the raised lake levels compared to the preceding and following periods (Kutzbach and Street-Perrott, 1985). However, this conceptual model does not fit the North American climate of 9,000 yr B.P., because of the residual effects of the melting ice sheet on climate (COHMAP Members, 1988).

The glacial and postglacial climate of North America can be viewed as a combination of the direct response to changes in solar radiation and the lingering effects of the slowly melting ice sheets. Several concepts have emerged from our detailed modeling studies (Kutzbach and Guetter, 1986; COHMAP Members, 1988; Kutzbach and Wright, 1985; Kutzbach 1987; and Webb *et al.*, 1987) and can be illustrated in the form of a vertically stacked set of "snapshots" of the climatic patterns from 18,000 yr B.P. to today (Fig. 10).

The Laurentide ice sheet, with a maximum elevation of around 3,000 m, had a profound effect on the glacial-age circulation (Fig. 10, bottom). A classical glacial anticyclone was produced by the air in contact with ice. This air was, of course, cold compared to the surrounding regions so that there was sinking motion above the ice sheet, low-level outflow of air along the ice perimeter and, via earth-rotational effects, a clockwise anticyclonic surface-wind regime (Hobbs, 1926; Brooks, 1926; Bryson and Wendland, 1967). A strong north-south temperature gradient developed near and south of the ice front and, above this southward-displaced polar front (southward-displaced compared to present), a major jet stream was located. Simultaneously, the northern flank of the ice sheet became a boundary between truly Arctic air masses and the relatively warmer (and sinking) air over the ice sheet itself. This secondary north-south thermal gradient was also associated with a mid-troposphere wind maximum. The glacial-age jet-stream pattern over North America could therefore be described as split into two branches (northern and southern) in contrast to the dominant single-jet core at present. (A more complete explanation of the split jet of glacial times involves dynamical processes related to flow around a mechanical barrier as well as the thermal process already mentioned.)

To the climatologist, the sketch of upper and low-level flow patterns is shorthand for many climatic concepts regarding thermal and moisture regimes: (1) storm-track precipitation in the southwest and along the south and east; (2) cold and dry conditions in the northwest; but (3) perhaps no conditions colder than at present in parts of Alaska where the flow had a more southerly component than at present. The cold northwesterly flow along the northeastern flank of the

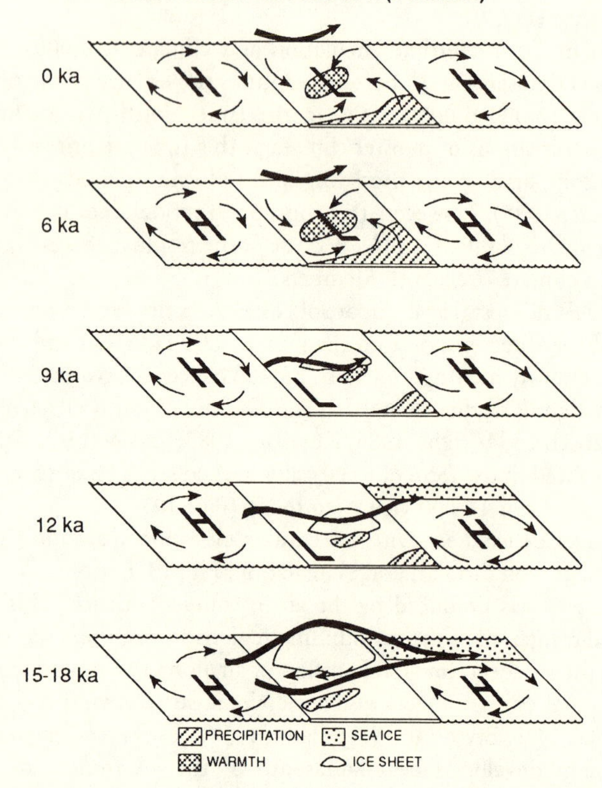

Fig. 10. Conceptual changes in summertime circulation for North America, 18 ka to present, based on simulation results of COHMAP Members (1988) and Figs. 5 through 9. At 15–18 ka the circulation was strongly influenced by the presence of the large, North-American ice sheet and by extensive sea ice in the North Atlantic. The jet stream split into two branches and extended far downstream over the Atlantic. The southern branch was shifted far south and was accompanied by precipitation associated with frequent cyclonic storms. Along the southern ice-sheet margin, conditions were relatively dry and dominated by a glacial anticyclone. By 12 ka the ice sheet was smaller and the sea ice less extensive. The jet stream was weaker than at 15–18 ka and was located along the southern ice margin; it was accompanied by precipitation associated with cyclonic storms. Increased summertime solar radiation, associated with orbital changes, helped establish warmer conditions west of the ice sheet, a low-pressure center in the southwest, and increased precipitation in the south and southeast. By 9 ka the ice sheet was very small and the orbitally produced enhancement of summertime conditions was becoming very evident. By 6 ka, northern summer monsoons over land and anticylonic circulations over the ocean, caused by orbitally induced changes in solar radiation, were stronger than at 9 ka (when the presence of the small ice sheet still moderated the orbital effects) and also stronger than at present (see Figs. 5–9). The interior of the continent was warmer than at present and the region of increased moisture extended farther into the continent on the south and east margins. At present, 0 ka, these summertime circulation features are not as pronounced as they were at 6 ka. (With permission: the AAAS; copyright 1988 the AAAS.)

ice sheet may have helped chill the North Atlantic and displace the sea-ice border southward (Manabe and Broccoli, 1985).

By 12,000 yr B.P., our modeling experiments point toward significant evolution of the climatic flow patterns (Fig. 10). The insolation increase during northern summer (Figs. 4, 7) had significantly warmed the region well south and west of the ice sheet, but regions near and downwind of the ice remained cold. Significant wastage of the North American ice sheet had reduced its extent and height (and perhaps reflectivity) sufficiently so that the glacial anticyclone had weakened, and the primary thermal gradient (and jet) hugged the southern boundary of the ice sheet. This change resulted in a single jet stream rather than the split jet of full-glacial time. In the east, the climate was cool and moist along the ice front. West of the ice sheet in parts of western North America, this time and the early Holocene were warm and dry, coinciding with maximum insolation (Ritchie *et al.*, 1983). The continuing cool conditions in central North America interfered with the establishment of the mid-continent heat low that had developed in North Africa/Eurasia (Fig. 9). Instead, our model simulations show only a hint of this pure monsoonal response. It is expressed as a center of lower pressure (than at present) and increased inflow to the southwest. Thus, the model suggests the possibility that in parts of the southwest the cool-moist glacial conditions could have given way to continued moister-than-present conditions associated with an enhanced summer monsoon.

The trends mentioned above for 12,000 yr B.P. became even more evident at 9,000 yr B.P. as the climatic effects of the ice sheet diminished further. By 6,000 yr B.P. the residual ice sheet was too small to fill even a single model grid-square, and therefore our simulations for North America for 6,000 yr B.P. show the simple monsoonal response to the increased insolation as occurred earlier in North Africa/Eurasia (Fig. 10). At 6,000 yr B.P., the maximum warming (compared to present) was centered in the continental interior. The enhanced monsoonal circulation corresponded closely to the pattern over North Africa/Eurasia (Fig. 9) except for being subdued because of the smaller size of the continent. At this time, the east and south may have experienced some increased summer precipitation because of the enhanced southerly flow. Some of the central interior, however, may have been drier owing to the increased evaporative losses associated with increased insolation and temperature and possibly decreased precipitation if the flow became more westerly to the west of the low-pressure center. This simulated climatic sequence (and the conceptual sketch, Fig. 10) perhaps comes closest to depicting and explaining why the mid-Holocene was the warmest and driest period in the North American continental interior (i.e., the Hypsithermal prairie expansion—see Wright, 1976a). Lake-level evidence from northern middle latitudes (Street-Perrott, 1986; Winkler *et al.*, 1986; Harrison, 1989) suggests that the climate of the period 9,000 to 6,000 yr B.P. was indeed drier than that before or after (Fig. 11). The simulated wintertime climate of the period around 9,000 to 6,000 yr B.P., when insolation was less than present, was generally colder than at present; however, this was modified by advective effects from the oceans in western North America (Kutzbach and Gallimore, 1988).

In summary, the glacial-age circulation and climate of North America was vastly different from present, as exemplified by the split jet and altered storm tracks (Fig. 10). The mid-Holocene warm and dry period, described by Wright (1968, 1976a) for the North-American Midwest, and simultaneously the moist phase in the northern tropics (Kutzbach and Street-Perrott, 1985; Fig. 11) were produced by the orbital change that increased summer warmth and reduced P-E in northern, mid-latitude continental interiors and increased P-E in the northern tropics. The climatic changes associated with orbital changes were somewhat larger in magnitude in North Africa/Eurasia compared to North America because the former region is larger in size.

Time series of temperature and precipitation at specific locations (Fig. 12) derived from our simulation studies (Webb *et al.*, 1987) show the combined effects of insolation and ice sheet on regional warmth and moisture. These time series provide a picture of the evolving climate in terms of standard climatic variables for comparison with standard paleoclimatic data derived from geological observations (next section).

Concepts from COHMAP Data

The combination of different types of paleoclimatic data into a global data set (Fig. 3) allowed COHMAP researchers to look at the global patterns of the climate changes over the past 18,000 years (COHMAP Members, 1988). We have chosen to use pollen and lake-level data from eastern North America to illustrate many of the lessons and concepts that are also derivable from the complete data set. We have focused on describing a zoom-lens view of data in order to illustrate how the traditional work by paleoecologists at sites is linked to the global and continental syntheses that have been a hallmark of COHMAP research.

Zoom-Lens View of Data and Its Implications

General Viewpoint

The 20,000-year record of mid-continental vegetation and climate can be viewed at several spatial scales covering areas of less than 1 km^2 up to almost 10^8 km^2 (Fig. 13). Such a perspective gives a zoom-lens view of the vegetation and the data. It allows examination of vegetational change at different levels of spatial resolution. Local changes can be placed within the context of changes that affect whole landscapes or regions, and the regional consequences of global-scale

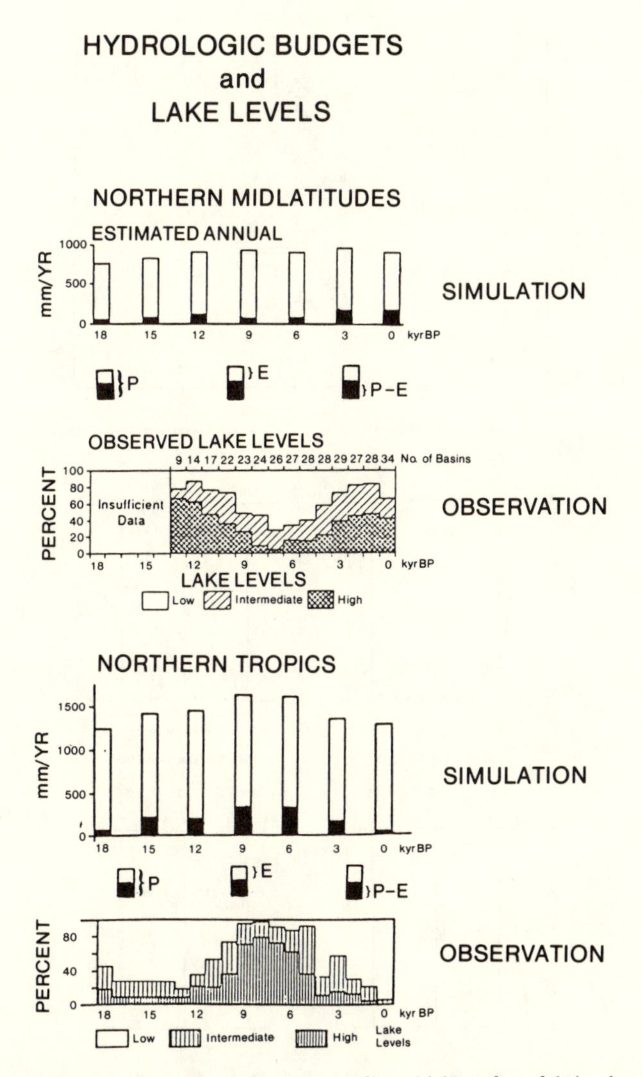

Fig. 11. Observed status of lake levels (low, intermediate, high) and model-simulated moisture budgets (precipitation, evaporation, and precipitation-minus-evaporation) for each 1,000 years (observations) or each 3,000-year interval (model) from 18,000 years ago to present. The temporal variations in percentage of lakes with low, intermediate, or high levels is taken from the Oxford lake-level data bank. In both northern midlatitudes (top) and the northern tropics (bottom), the shape of the high-lake-level curve corresponds roughly to the shape of the P-E curve. That is, low lakes and low P-E around 9–6 ka in northern midlatitudes and high lakes and high P-E occurred around 9–6 ka in the northern tropics. Modified from Kutzbach and Street-Perrott (1985) and Street-Perrott (1986). (Reprinted with permission: *Nature* 317: 133 and the Office for Interdisciplinary Earth Studies (OIES), University Corporation for Atmospheric Research (UCAR).)

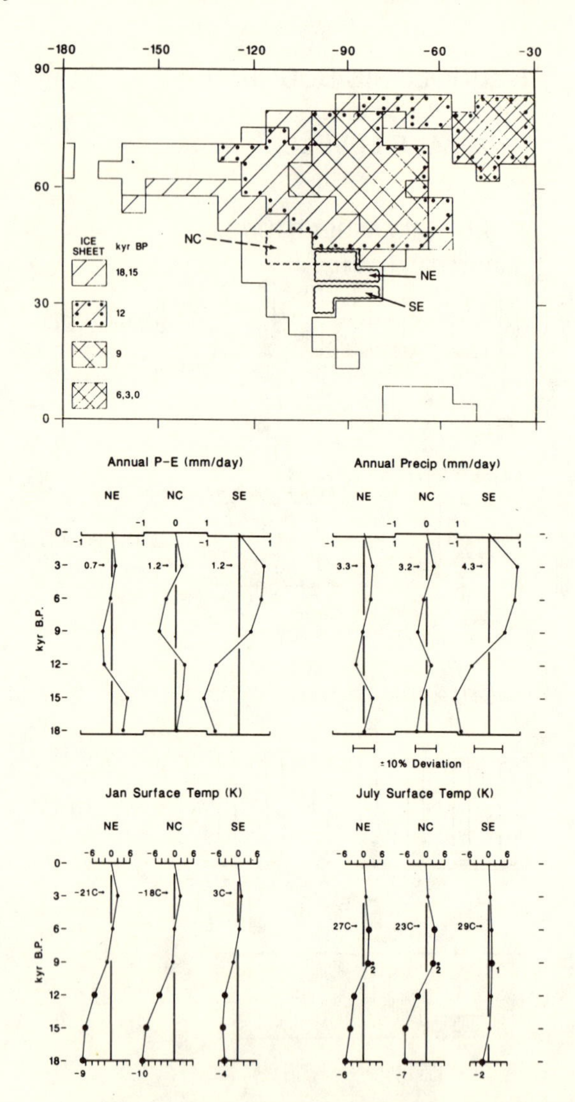

Fig. 12. Climatic values simulated by the NCAR Community Climate Model. Area-average changes of January and July surface temperature, annual precipitation, and precipitation-minus-evaporation (P-E) for three regions in eastern North America. The time series are departures from the control (0 ka) simulations. For annual precipitation, plus-or-minus 10% deviations from the control values are shown. The control value is written beside each series. Large dots indicate that the departure is significant, above the 95% confidence level, compared to the model's inherent variability. The unconnected circle and temperature value at 9 ka show the results of the "no ice sheet" experiment of Kutzbach and Guetter (1986). Figure from Webb *et al.* (1987) as modified from Kutzbach (1987) and reprinted with permission from the Geological Society of America, Decade of North American Geology series, Vol. K-3: 456.

variations can also be identified. Such hierarchical comparison of the data can prevent unique local changes from being mistaken for a global signal, and it should also make clear that globally forced variations produce markedly different changes among regions at a subcontinental scale (Kutzbach, 1976; Bartlein, 1988).

Contour maps of pollen percentages (called isopoll maps) play a key role in this zoom-lens view of the vegetation. Contours represent phenomena as if they were continuous and wavelike, whereas the vegetation at local sites and even within landscape mosaics can appear photon- or particle-like because it often differs markedly among isolated points and adjacent patches. Even the temporal changes at the local and landscape scales are sometimes independent of the apparently continuous changes at regional and subcontinental scales. Contours (and their implied abundance surfaces) therefore represent models that may explain most of the variance at the regional and subcontinental scales but may misrepresent the specific nature of changes at local sites. The zoom-lens perspective allows an investigator to appreciate the proper roles of contour maps and local pollen diagrams. It therefore accommodates both the broad-scale and local views of the data and permits resolution of such potential debates as whether firebreaks or climate allowed forest to replace prairie in Minnesota (Grimm, 1983). Each explanation is appropriate for a selected scale and for an appropriate amount of smoothing and detail at that scale (Webb *et al.*, 1983).

Specific Examples

The data at each spatial scale provide a unique view of the types, timing, and magnitude of changes during the past 20,000 years. At the finest scale of resolution (< 1 km^2), stratigraphic profiles of plant macrofossils and lowland pollen from single sites in Minnesota illustrate vegetational and environmental changes within the basin in which the data were collected (Figs. 14, 15). At the next step up, standard pollen diagrams of the upland pollen taxa illustrate the vegetational changes within a landscape and have a spatial resolution of about 1,000 km^2 (Fig. 15). Isopoll maps of the pollen data (Figs. 16, 17) from multiple sites provide even-broader-scale views of the changing vegetation at both the regional (10^5 to 10^6 km^2) and subcontinental scales (10^7 km^2), and COHMAP Members (1988) have published maps of vegetational changes that are as nearly global-scale as can currently be assembled (Figs. 3, 5, 13).

The diagrams of plant macrofossils (Fig. 14), wetland pollen, and diatoms indicate water-level changes in the immediate basin, and also indicate the changing vegetation and wetland conditions immediately surrounding the basin. At Kirchner Marsh, an ice block was buried at the site for 3,000 years after glacial retreat (Florin and Wright, 1969). Water levels began high, but they became low and fluctuated from 7,500 to 3,500 yr B.P. (Watts and Winter, 1966). After 2,000 yr B.P., a marsh developed in the basin (Wright *et al.*, 1963). These

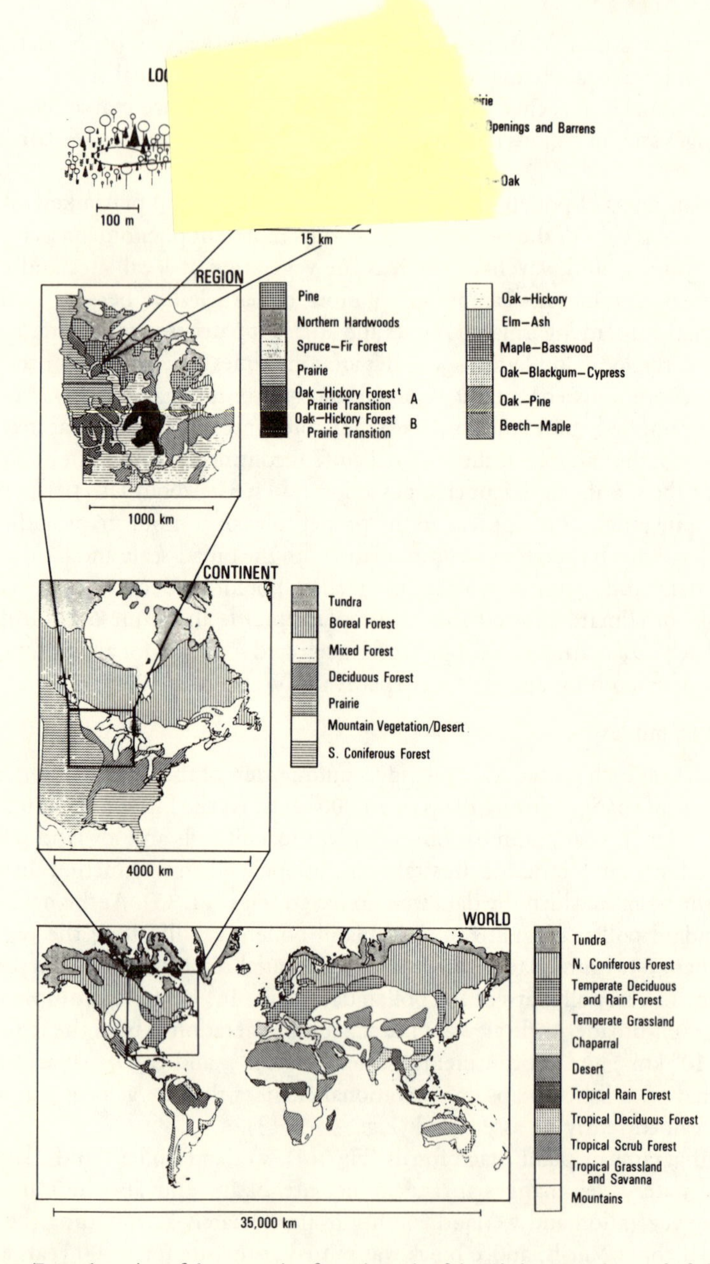

Fig. 13. Zoom-lens view of the vegetation from the scale of the Kirchner Marsh record of plant macrofossils up to the global scale. The local map is schematic; the landscape map is from Cushing (personal communication); the region map is modified from Mertz (1979); the continental map is modified from Webb (1988); and the world map is modified from Laporte (1968).

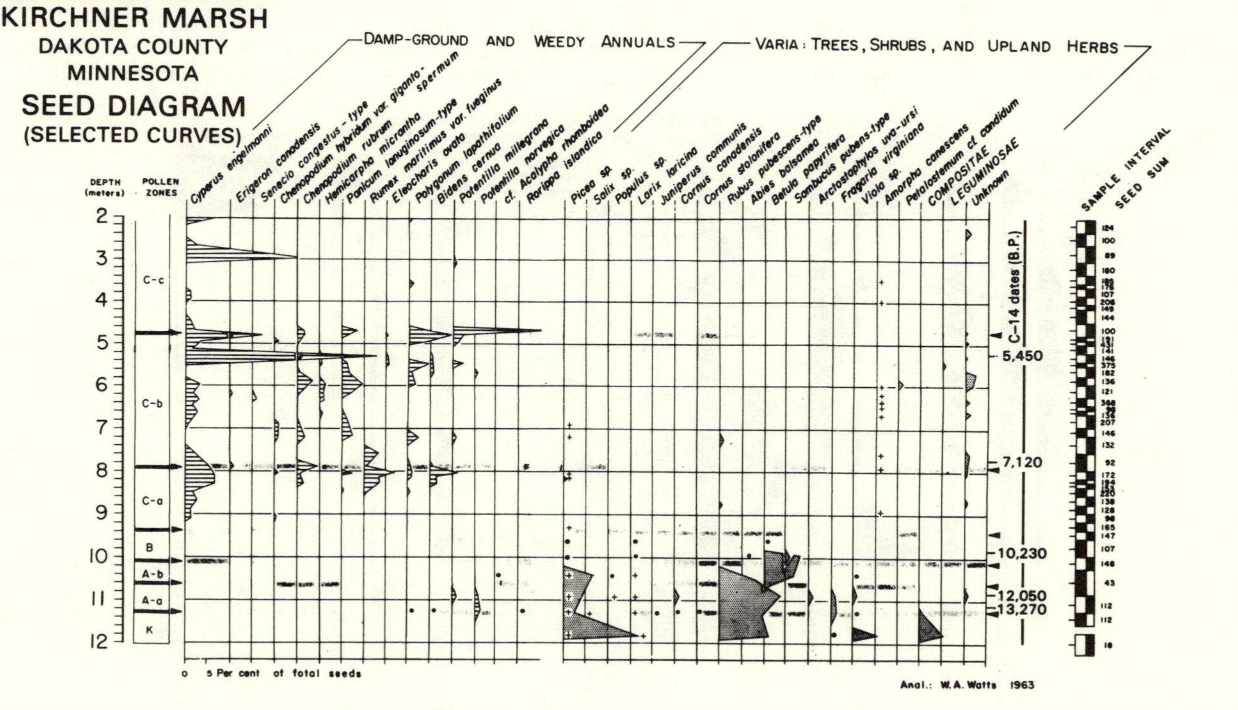

Fig. 14. Selected time series from the plant macrofossil diagram from Kirchner Marsh, Minnesota (after Watts and Winter, 1966). Time series represent percent of total seeds with seed sum listed to the *right* of the diagram. Most of the seeds are aquatics (not shown here) and damp-ground perennials, but seeds of boreal trees (spruce, *Picea*; birch, *Betula*) are most abundant before 10 ka and the seeds of annuals are most abundant after 9 ka and during the prairie period when water levels were lower than earlier (Watts and Winter, 1966). A lack of seeds in the sedge peat in the upper 2 m of the core precluded plotting the data during marsh development. (With permission: *Geological Society of America Bulletin* 77: attached plate.)

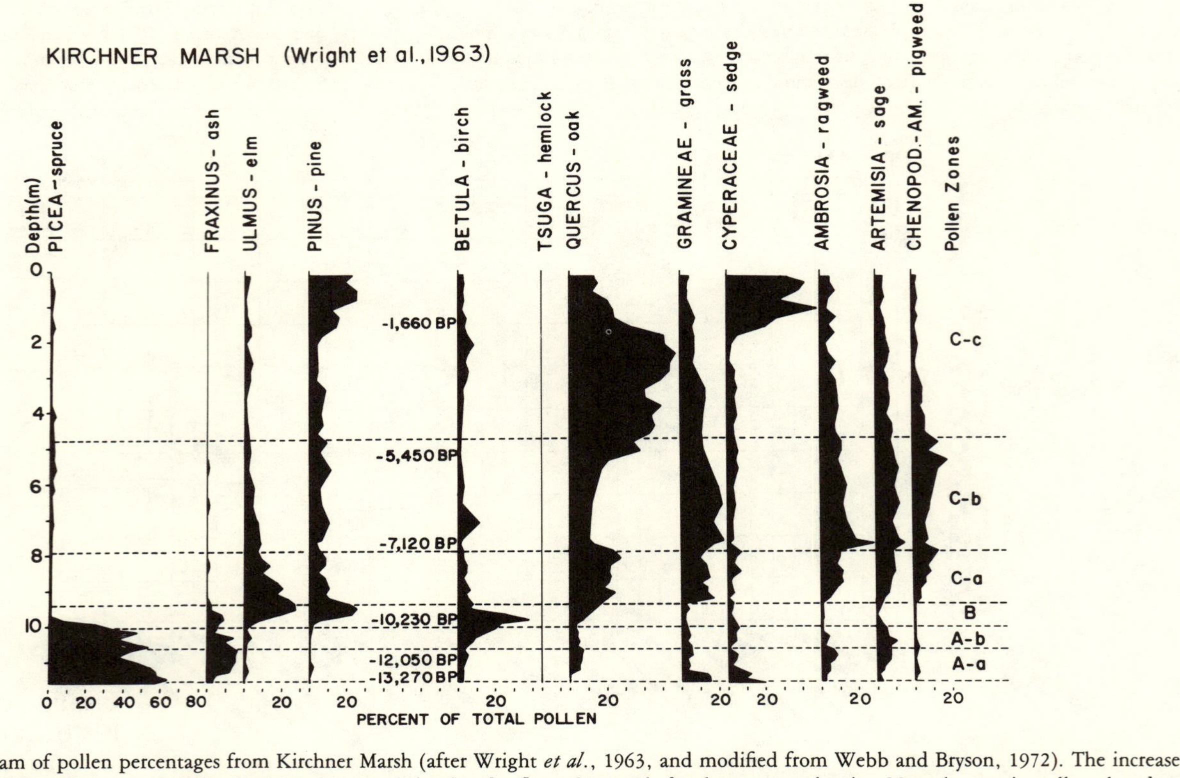

Fig. 15. Diagram of pollen percentages from Kirchner Marsh (after Wright *et al.*, 1963, and modified from Webb and Bryson, 1972). The increased abundance of sedge (Cyperaceae) pollen in the upper 2 m is local and reflects the marsh development at the site. Most changes in pollen abundance below 2 m reflect vegetational changes on the surrounding landscape. (With permission: *Quaternary Research*.)

local changes at the site occurred within the context of the broader-scale changes in the vegetation within the landscape (ca. 1,000 km^2) around this site. The pollen diagram of upland pollen types records these changes and shows that a spruce-dominated parkland and forest was replaced briefly by pine, alder, and birch before oak-scrub vegetation developed at 9,000 yr B.P. (Fig. 15). Prairie invaded about 7,500 yr B.P. and was replaced after 3,500 yr B.P. by oak forest that continues to grow around the site. In general, the relative moisture and associated vegetation within the basin changed in parallel with the landscape vegetation.

Maps of changes in the regional vegetation show how both the local changes at the site and the vegetational changes on the landscape fit in with regional changes in the vegetation. Spruce trees became abundant first in the south and later in the north between 12,000 and 8,000 yr B.P. (Fig. 16). Pine trees entered Minnesota from the east about 10,000 yr B.P., and oaks moved north after 11,000 yr B.P. The prairie formed just after 10,000 yr B.P., then moved eastward and later westward (Fig. 16). Water levels at Kirchner Marsh were lowest when the prairie was most prevalent in east-central Minnesota (Watts and Winter, 1966). The local and landscape changes at Kirchner Marsh therefore reflected these regional changes that themselves were part of subcontinental and global-scale changes (Fig. 16).

The maps for the major taxa at the subcontinental scale (Fig. 17) show that the prairie/forest movements that so dominate the record at Kirchner Marsh are small in comparison with the general sweep of vegetational changes across the continent. Here the changing abundance, distribution, and association among the plant taxa led to continuous changes in the vegetation with plant assemblages appearing, disappearing, reorganizing, and moving (Jacobson *et al.*, 1987; Webb, 1987, 1988). Between 18,000 and 10,000 yr B.P., the vegetation differed from that of today, and a spruce parkland grew over a broad region south of the ice sheets. Rather abruptly between 12,000 and 10,000 yr B.P., the spruce parkland disappeared (as the broad overlap between spruce and sedge pollen decreased), and several modern plant assemblages emerged including the deciduous forest and mixed forest (Figs. 16, 17). Later assemblages including the prairie, tundra, boreal forest, and southeast pine-oak forest appeared. These modern associations have existed since then, and most assemblages of fossil pollen are analogous to one or more samples of modern pollen—that is, they have had modern analogs (Fig. 18). Before 10,000 yr B.P., fewer than half the assemblages had modern analogs (Fig. 18).

During the entire period mapped (from 18,000 to 10,000 yr B.P.), and independent of whether many or few modern analogs existed, spatial patterns always were present in the vegetation. Vegetational changes have varied among regions, leading to synoptic patterns on the isopoll maps and to major differences among pollen stratigraphies from region to region. For instance, the

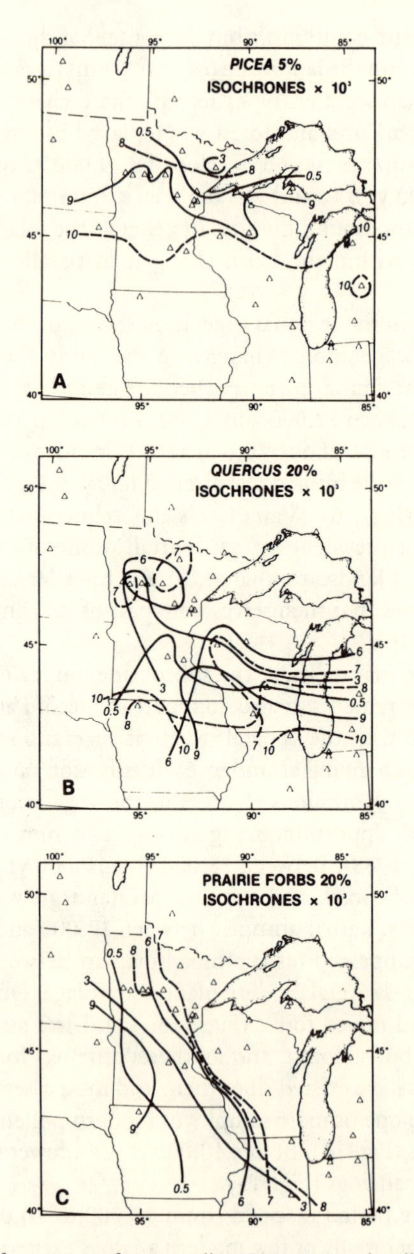

Fig. 16. Isochrone maps for contours of spruce pollen (*Picea*, 5%), oak pollen (*Quercus*, 20%), and prairie forbs pollen (*Artemisia*, other Compositae, Chenopodiaceae/Amaranthaceae, 20%) from 10 ka to 0.5 ka (from Bartlein *et al.*, 1984). Spruce values increased to the north, oak to the south, and prairie forbs to the west. (With permission: *Quaternary Research* 22: 364.)

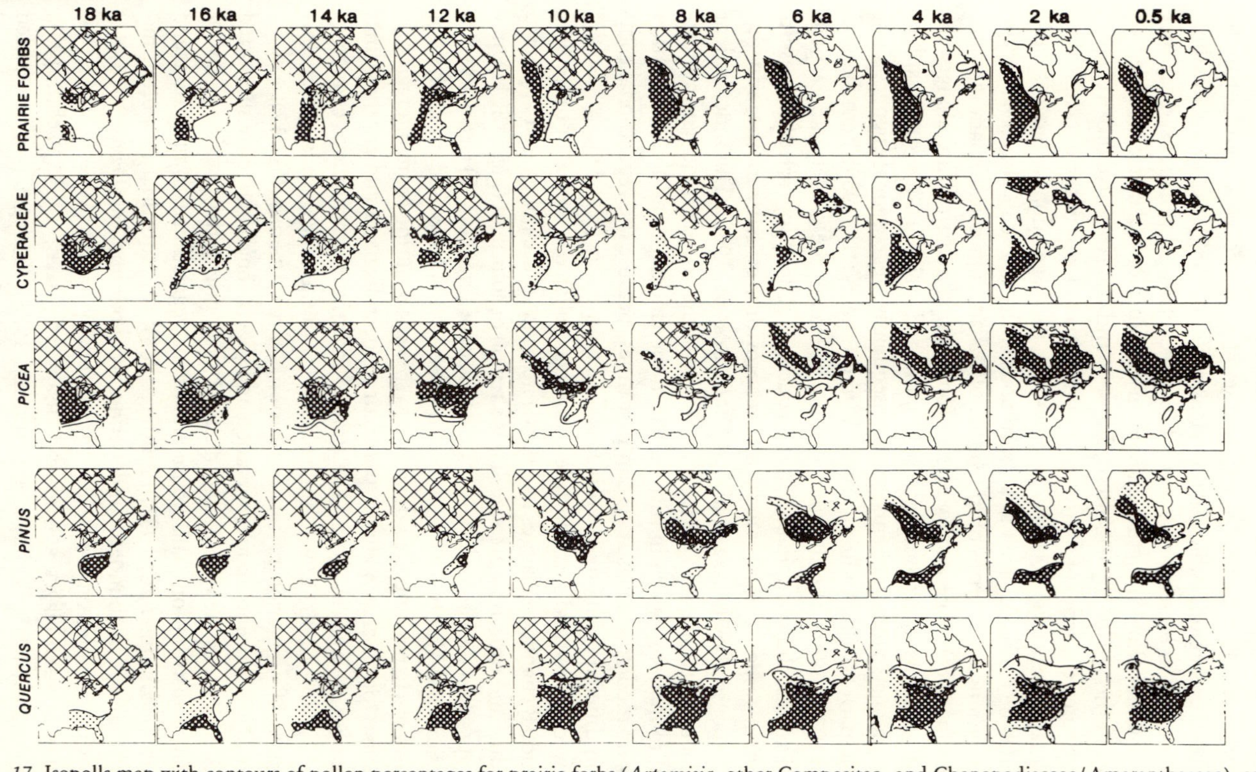

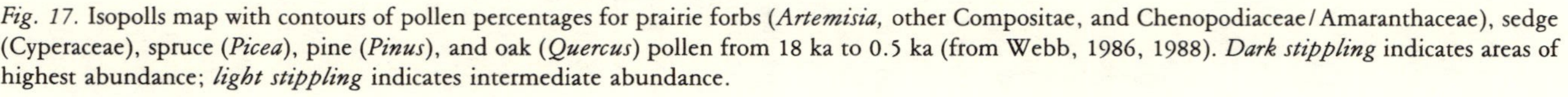

Fig. 17. Isopolls map with contours of pollen percentages for prairie forbs (*Artemisia,* other Compositae, and Chenopodiaceae / Amaranthaceae), sedge (Cyperaceae), spruce (*Picea*), pine (*Pinus*), and oak (*Quercus*) pollen from 18 ka to 0.5 ka (from Webb, 1986, 1988). *Dark stippling* indicates areas of highest abundance; *light stippling* indicates intermediate abundance.

movements of the prairie/forest border dominate in the western Midwest but are unimportant to the east. No standard set of pollen zones is therefore applicable above the regional scale (Gaudreau and Webb, 1985). COHMAP Members (1988) provide an even broader view of these vegetational variations. They present maps that show how the changes in eastern North America are tied to changes in the rest of the continent, Europe, and the North Atlantic (Fig. 5).

This zoom-lens series of figures, from diagrams to maps of the data at different spatial scales (Figs. 5, 12–17), shows that the vegetation at broad spatial scales may be conceptualized as a continuous entity whose local and landscape representations are in terms of individual plants and a mosaic of plant taxa and communities. The vegetation varies in composition and structure along well-defined gradients at regional and subcontinental scales. Pollen stratigraphies also vary along spatial gradients, and consequently, vegetational changes show synoptic patterns (McAndrews, 1966; Wright and Watts, 1969; Bernabo and Webb, 1977). When the abundance of pollen taxa in the modern landscape is plotted against climatic variables in the same spatial domain, each of the resulting climatic-response surfaces is distinctive, providing evidence for the individualistic response of plant taxa to climate (Bartlein *et al.*, 1986). This independent behavior of individual taxa leads through time to the disappearance and appearance of formations, which in turn results in the landscape and local vegetation having no modern analogs at certain past times, because the vegetational composition differed from any assemblages that grow today (Fig. 18).

Viewed in time and space at different scales, pollen data reveal both steplike and gradual changes. Changes like the decrease in spruce pollen that appear to be gradual and time-transgressive across a region or continent (Figs. 16, 17) may be abrupt when they occur locally or on a landscape (Figs. 14, 15) (see discussion in Bernabo and Webb, 1977). Jacobson *et al.* (1987) have used an analysis of the times-of-change in pollen diagrams to show how during some intervals major changes for different pollen types occur in most diagrams and at other times the changes are asynchronous. The time of maximum synchronous changes is about 11,000 ± 2,000 yr B.P. and coincides with the time of most major rearrangement of vegetation patterns in eastern North America (Fig. 17).

Several key concepts about vegetation and vegetational change are supported by this zoom-lens view of pollen data: (1) synoptic patterns exist in taxon-abundance distributions and consequently in vegetational composition and structure (Jacobson *et al.*, 1987; Huntley, 1988); (2) individual taxa often migrate and change in abundance independently of one another (Davis, 1976); (3) pollen stratigraphies can be locally and regionally unique (Wright, 1968); (4) vegetational assemblages appear and disappear (Davis, 1983; Webb, 1987; Jacobson *et al.*, 1987; Webb, 1988; Huntley, 1988); (5) vegetation has existed that had no modern analogs (Wright *et al.*, 1963; Overpeck *et al.*, 1985); and (6) vegetational changes are sometimes gradual and sometimes abrupt (Webb,

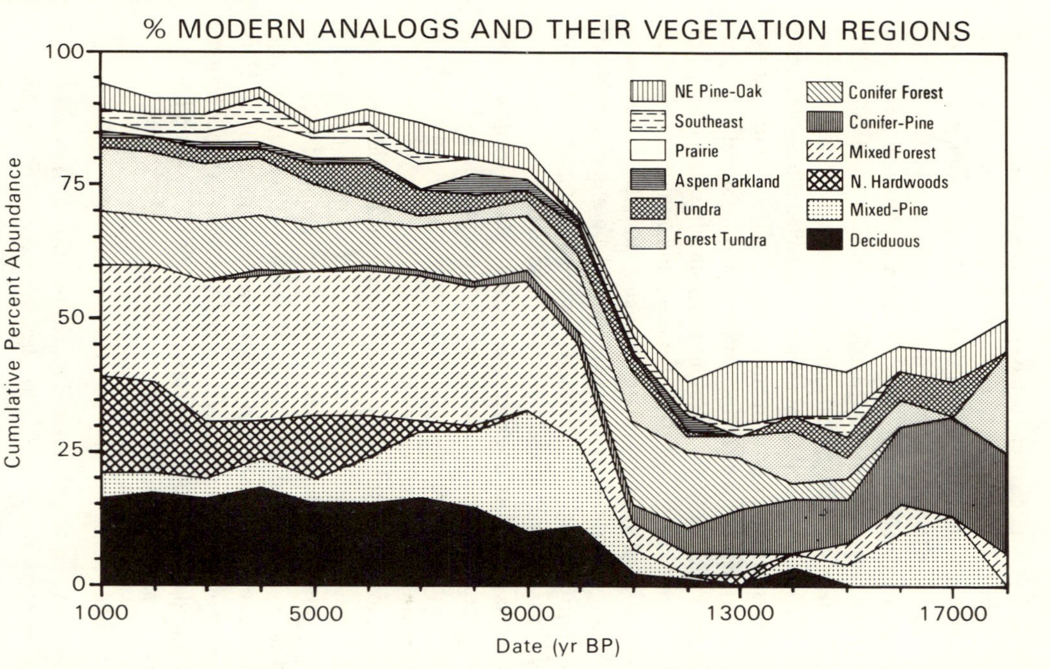

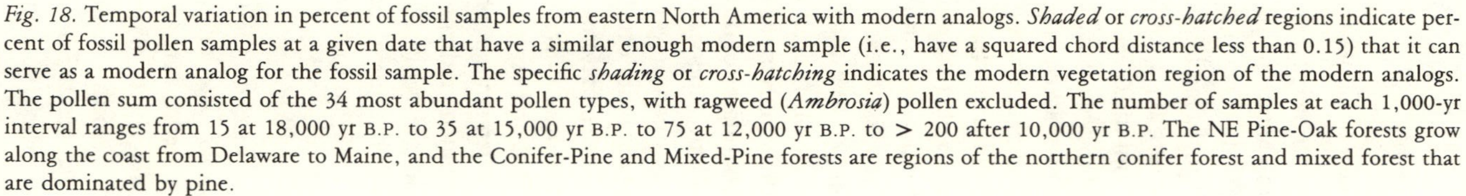

Fig. 18. Temporal variation in percent of fossil samples from eastern North America with modern analogs. *Shaded* or *cross-hatched* regions indicate percent of fossil pollen samples at a given date that have a similar enough modern sample (i.e., have a squared chord distance less than 0.15) that it can serve as a modern analog for the fossil sample. The specific *shading* or *cross-hatching* indicates the modern vegetation region of the modern analogs. The pollen sum consisted of the 34 most abundant pollen types, with ragweed (*Ambrosia*) pollen excluded. The number of samples at each 1,000-yr interval ranges from 15 at 18,000 yr B.P. to 35 at 15,000 yr B.P. to 75 at 12,000 yr B.P. to > 200 after 10,000 yr B.P. The NE Pine-Oak forests grow along the coast from Delaware to Maine, and the Conifer-Pine and Mixed-Pine forests are regions of the northern conifer forest and mixed forest that are dominated by pine.

1988) or may be time-transgressive regionally but abrupt locally (Bernabo and Webb, 1977; Webb *et al.*, 1983).

Ties to Climatic Change

Knowledge of how the climate system behaves and how vegetation responds to climate helps explain this diversity of vegetational patterns and variations (Webb, 1986). It is regional to local climates that affect vegetation at a site or across a landscape, but these regional climatic patterns are influenced by factors acting on a continental to global scale. The top-down understanding gained from the COHMAP modeling experiments (Kutzbach and Guetter, 1986; COHMAP Members, 1988) shows that the global climate system has spatial gradients well marked in several climate variables that change in correlated but sometimes independent ways across eastern North America (Fig. 12). These gradients and conditions result in spatial patterns of all the climate variables as well as in patterns of changes in the variables. Not all regions show simultaneous changes of similar magnitude or even direction for temperature and moisture (Fig. 12). New combinations of climate variables create no-analog climates (i.e., climates unlike any today). The continuously changing, global boundary conditions (Figs. 4, 6) lead to both gradual and abrupt vegetational changes, because of thresholds in the way that vegetation and climate respond to slow changes in the boundary conditions (e.g., the jet is either split or not split in its flow across North America, depending on the ice-sheet size and height, and the climate conditions favoring abundant growth of a taxon may decrease markedly in favorability over a narrow temperature range).

Such patterns of regional climate change, coupled with the individualistic behavior of different plant taxa (Bartlein *et al.*, 1986), led to the appearance of local and regional vegetation without modern analogs, to synoptic patterns in pollen stratigraphies, to abrupt and gradual vegetational changes, and to well-defined gradients in taxon abundances. Local variations in some plant distributions also arise because of short-distance variation in soils or disturbance regimes and because of local wetland development (Figs. 13, 14). See Webb and Bartlein (1988) and Prentice *et al.* (in press) for further discussion.

Ties to Lake-Level Data

Tests for the ties between vegetation and climate require information from other data that can be dated and mapped. Street and Grove (1976) first demonstrated that changing water levels in lakes show synoptic patterns. Harrison and Metcalfe (1985) and Winkler *et al.* (1986) have recently summarized the data for North America and the northern Midwest and illustrated how the local changes at Kirchner Marsh and Lake Carlson (Wright *et al.*, 1963; Watts and Winter, 1966) fit with the general patterns of regional and continental vegetation and climatic change. The patterns in lake-level variations fit in well with those climatic and vegetational changes inferred from pollen data (R. Webb, 1990; T. Webb *et al.*, in press). For example, the water levels in mid-continent lakes were generally low from 7,000 to 3,000 yr B.P. (Fig. 11), at the same time that the prairie forest border extended farthest east (Street-Perrott, 1986; Webb *et al.*, 1983, in press; Harrison, 1989).

Discussion and Conclusions

This book honors H. E. Wright, Jr., who was asking and answering climatic questions of what? when? and why? before COHMAP developed. He helped establish COHMAP as an interdisciplinary project aimed at quantitative analysis of geologic observations and testing of climatic hypotheses with numerical climate models.

As our brief historical review has illustrated, we started with a good foundation. The ideas that orbital changes paced ice ages and monsoons were advanced by Milankovitch, Spitaler, Hays *et al.* (1976), and many others both before and after COHMAP research began. Climatological analyses of Köppen, Wegener, and Brooks clearly indicated that both direct orbital effects and lagged effects of melting ice sheets need to be considered in answering the major climatic puzzles of the glacial-to-interglacial march of global climate.

COHMAP research has contributed significantly and quantitatively to the solution of some of these puzzles of Holocene climate change (COHMAP Members, 1988). Thanks to the quantitative and accurate analyses now possible in dating, in climatic interpretation of geologic observations, and in numerical climate simulation, COHMAP has clarified key aspects of the response of late Quaternary climate to orbitally induced solar-radiation variations and to the lingering effects of the glacial age (ice sheets, cold oceans, lowered atmospheric carbon-dioxide concentration, and so on). COHMAP research has produced a conceptual framework, as summarized in this chapter, for understanding the reasons for mid-Holocene growing-season warmth and dryness in North America (and central Asia); for the simultaneous moist phase in northern, tropical monsoon lands; for the occurrence of somewhat larger, Holocene climatic changes in North Africa/Eurasia compared to North America; and for the lag of certain climatic changes in eastern North America. It has also provided a plausible explanation for past vegetation having no modern analogs and for major vegetational regions appearing and disappearing.

In addition to helping answer these climatic and vegetational puzzles of the late Quaternary, COHMAP research has helped to provide a framework for addressing other questions of biosphere dynamics (Webb, 1986, 1988; Prentice *et al.*, in press, Huntley and Webb, 1989). It also has helped to test models that are being developed to estimate future climates (COHMAP Members, 1988).

The zoom-lens perspective developed in this chapter helps to show how the local and regional changes recorded at sites sampled for pollen (e.g., Kirchner Marsh) fit into the broad-scale climate and vegetation changes mapped and modeled by COHMAP. The latter changes will never explain all the variability at individual sites, but investigators can use this perspective to sort out the spectrum of factors inducing change at these sites. Explanations can then be developed to allow for the differing role and magnitude of biotic factors, fire, and climate at each spatial scale.

Acknowledgments

Grants from the National Science Foundation (NSF) (Climate Dynamics Program) and the Department of Energy (Carbon Dioxide Research Division) provided the major support for this project. Additional support for COHMAP was provided by the Nuclear Regulatory Commission to NSF. The specific grant numbers are: NSF/ATM-8713981 and DOE FG02-85ER60304 (Brown University) and NSF/ATM-8603295 (University of Wisconsin-Madison). The numerical climate-model simulations were made at the NSF-sponsored National Center for Atmospheric Research, Boulder, Colo., with computing grant 35381017. For help in this paper, we thank K. Anderson, P. J. Bartlein, P. Behling, M. Kennedy, F. A. Street-Perrott, R. S. Webb, and M. Woodworth. For help and encouragement, we thank R. A. Bryson, P. J. Bartlein, P. Behling, R. Chervin, K. Gajewski, R. Gallimore, D. C. Gaudreau, E. C. Grimm, P. Guetter, S. Harrison, S. Hastenrath, S. E. Howe, B. Huntley, M. Kennedy, P. Klinkman, B. Molfino, P. Newby, B. Otto-Bliesner, J. T. Overpeck, W. Prell, I. C. Prentice, S. H. Schneider, L. C. K. Shane, F. A. Street-Perrott, R. Steventon, A. Swain, W. Washington, R. S. Webb, and M. Woodworth, who is the administrative coordinator for COHMAP, as well as the other 22 coauthors of COHMAP Members (1988). We would also like to thank many of the program managers of the various funding agencies who not only provided the mechanism for funding but also provided encouragement and constructive criticism: Gene Bierly (NSF), Alan Hecht (NSF), T. Crowley (NSF), J. Fein (NSF), H. Virji (NSF), W. Curry (NSF), M. Riches (DOE), and R. Kornasiewicz (NRC).

References

Adhémar, J. A. (1842). "Révolutions de la mer." Privately published, Paris.

Antevs, E. (1947). The Great Basin, with emphasis on glacial and postglacial times. *University of Utah Bulletin* **38**, 168–191.

Bartlein, P. J. (1988). Climate history. *In* "Vegetation History" (B. Huntley and T. Webb III, Eds.), pp. 113–152. Kluwer Academic Publishers, Dordrecht.

Bartlein, P. J., Prentice, I. C., and Webb, T. III (1986). Climatic response surfaces based on pollen from some eastern North America taxa. *Journal of Biogeography* **13**, 35–57.

Bartlein, P. J., and Webb, T. III (1985). Mean July temperature for eastern North America at 6,000 yr BP: Regression equations for estimates based on fossil-pollen data. *Syllogeus* **55**, 301–302.

Bartlein, P. J., Webb, T. III, and Fleri, E. (1984). Holocene climatic change in the northern Midwest: Pollen derived estimates. *Quaternary Research* **13**, 35–57.

Berger, A. L. (1978). Long-term variations of caloric solar radiation resulting from the earth's orbital elements. *Quaternary Research* **9**, 139–167.

Bernabo, J. C., and Webb, T. III (1977). Changing patterns in the Holocene pollen record from northeastern North America; A mapped summary. *Quaternary Research* **8**, 64–96.

Broecker, W. S., Thurber, D. L., Goddard, J., Ku, T., Matthews, R. K., and Mesoella, K. J. (1968). Milankovitch hypothesis supported by precise dating. *Science* **159**, 1–4.

Brooks, C. E. P. (1922). "Evolution of Climate." Benn Brothers, London, 173 pp. (Reprinted in 1978, AMS Press, New York.)

——. (1926). "Climate Through the Ages; a study of the climatic factors and their variations." E. Benn Limited, London, 439 pp. (2nd rev. ed., 1949, Dover, New York, 395 pp.)

Bryson, R. A., and Wendland, W. M. (1967). Tentative climatic patterns for some late-glacial and postglacial episodes in central North America. *In* "Life, Land, and Water" (W. J. Mayer-Oakes, Ed.), pp. 271–289. University of Manitoba Press, Winnipeg.

CLIMAP (1976). The surface of the ice-age earth. *Science* **191**, 1138–1144.

——. (1981). Seasonal reconstructions of the earth's surface at the last glacial maximum. *Geological Society of America Map and Chart Series* MC-36.

COHMAP Members (1988). Climatic changes of the last 18,000 years: Observations and model simulations. *Science* **241**, 1043–1052.

Croll, J. (1864). On the physical cause of the change of climate during geological epochs. *Philosophical Magazine* **4**, 28, 121–137.

Crowley, T. J., Short, D. A., Mengle, J. G., and North, G. R. (1986). Seasonality in the evolution of climate during the last 100 million years. *Science* **231**, 579–584.

Davis, M. B. (1976). Pleistocene biogeography of temperature deciduous forests. *Geoscience and Man* **13**, 13–26.

——. (1983). Holocene vegetational history of the eastern United States. In "Late-Quaternary Environments of the United States," Vol. 2 (H. E. Wright, Jr., Ed.), pp. 166–181. University of Minnesota Press, Minneapolis.

Davis, M. B., Spear, R. W., and Shane, L. C. K. (1980). Holocene climates of New England. *Quaternary Research* **14**, 240–250.

Deevey, E. S., and Flint, R. F. (1957). Postglacial Hypsithermal interval. *Science* **125**, 182–184.

Emiliani, C. (1955). Pleistocene temperatures. *Journal of Geology* **63**, 538–578.

Florin, M.-B., and Wright, H. E., Jr. (1969). Diatom evidence for the persistence of stagnant glacial ice in Minnesota. *Geological Society of America* **80**, 695–704.

Gallimore, R. G., and Kutzbach, J. E. (1989). Effects of soil moisture on the sensitivity of a climate model to earth orbital forcing at 9000 yr BP. *Climatic Change* **14**, 175–205.

Gates, W. L. (1976). Modeling the ice-age climate. *Science* **191**, 1138–1144.

Gaudreau, D. C., and Webb, T. III (1985). Late Quaternary pollen stratigraphy and isochrone maps for the northeastern United States. *In* "Pollen Records of Late-Quaternary North American Sediments" (V. M. Bryant and R. G. Holloway, Eds.), pp. 247–280. American Association of Stratigraphic Palynologists Foundation, Dallas.

Grimm, E. C. (1983). Chronology and dynamics of vegetation change in the prairie-woodland region of southern Minnesota. *New Phytologist* **93**, 311–350.

Harrison, S. P. (1989). Lake levels and climatic change in eastern North America. *Climate Dynamics* **3**, 157–167.

Harrison, S. P., and Metcalfe, S. E. (1985). Spatial variations in lake levels since the last glacial maximum in the Americas north of the equator. *Zeitschrift für Gletscherkunde und Glazialgeologie* **21**, 1–15.

Hays, J. D., Imbrie, J., and Shackleton, N. J. (1976). Variations in the earth's orbit: Pacemaker of the ice ages. *Science* **194**, 1121–1131.

Hobbs, W. H. (1926). "The Glacial Anticyclone. The Poles of Atmospheric Circulation." Macmillian, New York.

Howe, S. E., and Webb, T. III (1983). Calibrating pollen data in climatic terms: Improving the method. *Quaternary Science Reviews* **2**, 17–51.

Huntley, B. (1988). Europe. *In* "Vegetation History" (B. Huntley and T. Webb III, Eds.), pp. 341–383. Kluwer Academic Publishers, Dordrecht.

Huntley, B., and Prentice, I. C. (1988). July temperatures in Europe from pollen data, 6000 years before present. *Science* **241**, 687–690.

Huntley, B., and Webb, T. III (1989). Migration: species' response to climatic variations caused by changes in the earth's orbit. *Journal of Biogeography* **16**, 5–19.

Imbrie, J., and Imbrie, K. P. (1979). "Ice Ages: Solving the Mystery." Enslow Publishers, Short Hills, N.J., 224 pp.

——. (1980). Modeling the climatic response to orbital parameter variations. *Science* **207**, 943–953.

Imbrie, J., and Kipp, N. G. (1971). A new micropaleontological method for quantitative paleoclimatology: Application to a Late Pleistocene Caribbean core. *In* "The Late Cenozoic Glacial Ages" (K. Turekian, Ed.), pp. 71–181. Yale University Press, New Haven.

Jacobson, G. L., Jr., Webb, T. III, and Grimm, E. C. (1987). Patterns and rates of vegetation change during the deglaciation of North America. *In* "The Geology of North America, Vol. K-3, North America and Adjacent Oceans during the Last Deglaciation" (W. F. Ruddiman and H. E. Wright, Jr., Eds.), Chapter 13, pp. 277–288. Geological Society of America, Boulder, Colo.

Kerr, R. A. (1984). Climate since the ice began to melt. Research News, *Science* **226**, 326–327.

——. (1986). Mapping orbital effects on climate. *Science* **234**, 383–384.

Köppen, W., and Wegener, A. (1924). Die Klimates des Quartärs. *In* "Die Klimate der Geologischen Vorzeit." Verlag von Gebrüder Borntraeger, Berlin.

Kutzbach, J. E. (1976). The nature of climate and climatic variations. *Quaternary Research* **6**, 471–480.

——. (1980). Estimates of past climate at Paleolake Chad, North Africa, based on a hydrological and energy-balance model. *Quaternary Research* **14**, 210–223.

——. (1981). Monsoon climate of the early Holocene: Climatic experiment using the Earth's orbital parameters for 9000 years ago. *Science* **214**, 59–61.

——. (1985). Modeling of paleoclimates. *In* "Advances in Geophysics," Vol. 28A (S. Manabe, Ed.), pp. 159–196. Academic Press, New York.

——. (1987). Model simulations of the climatic patterns during the deglaciation of North America. *In* "The Geology of North America, Vol. K-3, North America and Adjacent Oceans during the Last Deglaciation" (W. F. Ruddiman and H. E. Wright, Jr., Eds.), Ch. 19, pp. 425–446. Geological Society of America, Boulder, Colo.

Kutzbach, J. E., and Gallimore, R. G. (1988). Sensitivity of a coupled atmosphere/mixed layer ocean model to changes in orbital forcing at 9000 years B.P. *Journal of Geophysical Research* **93**, D1, 803–821.

Kutzbach, J. E., and Guetter, P. J. (1986). The influence of changing orbital parameters and surface boundary conditions on climate simulations for the past 18,000 years. *Journal of the Atmospheric Sciences* **43**, 1726–1759.

Kutzbach, J. E., and Otto-Bliesner, B. L. (1982). The sensitivity of the African-Asian monsoonal climate to orbital parameter changes for 9000 yr B.P. in a low-resolution general circulation model. *Journal of the Atmospheric Sciences* **39**, 1177–1188.

Kutzbach, J. E., and Street-Perrott, F. A. (1985). Milankovitch forcing of fluctuations in the level of tropical lakes from 18 to 0 kyr B.P. *Nature* **317**, 130–134.

Kutzbach, J. E., and Wright, H. E. Jr. (1985). Simulation of the climate of 18,000 yr BP: Results for the North American/North Atlantic/European sector and comparison with the geologic record. *Quaternary Science Reviews* **4**, 147–187.

Lamb, H. H., Lewis, R. P. W., and Woodroffe, A. (1966). Atmospheric circulation and the main climatic variables between 8000 and 0 B.C.: Meteorological evidence. *In* "World Climate from 8000 to 0 B.C." (T. S. Sawyer, Ed.), pp. 174–217. Royal Meteorological Society, London.

Lamb, H. H., and Woodroffe, A. (1970). Atmospheric circulation during the last ice age. *Quaternary Research* **1**, 29–58.

Laporte, L. F. (1968). "Ancient Environments." Prentice-Hall, Englewood Cliffs, N.J.

Manabe, S., and Broccoli, A. J. (1985). The influence of continental ice sheets on the climate of an ice age. *Journal of Geophysical Research* **90**, 2167–2190.

Manabe, S., and Hahn, D. G. (1977). Simulation of the tropical climate of an ice age. *Journal of Geophysical Research* **82**, 3889–3911.

McAndrews, J. H. (1966). Postglacial history of prairie, savanna, and forest in northeastern Minnesota. *Torrey Botanical Club Memoirs* **22**, 1–72.

Mertz, R. W. (1979). "Forest Atlas of the Midwest." U.S. Forest Service North Central Forest Experiment Station, St. Paul, Minn.

Mesoella, K. J., Matthews, R. K., Broecker, W. S., and Thurber, D. L. (1969). The astronomical theory of climatic change: Barbados data. *Journal of Geology* **77**, 250–274.

Milankovitch, M. (1920). "Théorie mathématique des phénomenes thermiques produits par la radiation solaire." Gauthier-Villars, Paris, 338 pp.

Mitchell, J. M. (1965). Theoretical paleoclimatology. *In* "The Quaternary of the United States" (H. E. Wright, Jr. and D. G. Frey, Eds.). Princeton University Press, Princeton, N.J.

Overpeck, J. T., Webb, T. III, and Prentice, I. C. (1985). Quantitative interpretation of fossil pollen spectra: dissimilarity coefficients and the method of modern analogs. *Quaternary Research* **23**, 87–108.

Peterson, G. M., Webb, T. III, Kutzbach, J. E., van der Hammen, T., Wijmstra, T. A., and Street, F. A. (1979). The continental record of environmental conditions at 18,000 yr B.P.: An initial evaluation. *Quaternary Research* **12**, 47–82.

Pokras, E. M., and Mix, A. C. (1985). Eolian evidence for spatial variability of late Quaternary climates in tropical Africa. *Quaternary Research* **24**, 137–149.

Prell, W. L. (1984a). Variation of monsoonal upwelling; A response to changing solar radiation. *In* "Climate Processes and Climate Sensitivity," *Geophysical Monograph Series* 29, American Geophysical Union, Washington, D.C.

——. (1984b). Monsoonal climate of the Arabian Sea during the late Quaternary; A response to changing solar radiation. *In* "Milankovitch and Climate" (A. Berger, J. Imbrie, J. Hays, G. Kukla, and B. Saltzmann, Eds.), pp. 349–366. D. Reidel, Hingham, Mass.

Prell, W. L., and Kutzbach, J. E. (1987). Monsoon variability over the past 150,000 years. *Journal of Geophysical Research* **92**(D7), 8411–8425.

Prentice, I. C., Bartlein, P. J., and Webb, T. III (in press). Vegetation change in eastern North America since the last glacial maximum: a response to continuous climatic forcing. *Journal of Ecology*.

Ritchie, J. C., Cwynar, L. C., and Spear, R. W. (1983). Evidence from north-west Canada for an early Holocene Milankovitch thermal maximum. *Nature* **305**, 126–128.

Rossignol-Strick, M. (1983). African monsoons, an immediate climate response to orbital insolation. *Nature* **303**, 46–49.

Saltzman, B. (1985). Paleoclimatic modeling. *In* "Paleoclimate Analysis and Modeling" (A.D. Hecht, Ed.), pp. 341–396. Wiley, New York.

Sears, (1942). Xerothermic theory. *Botanical Review* **8**, 709–739.

Shackleton, N. J., and Opdyke, N. D. (1973). Oxygen isotope and paleomagnetic stratigraphy of equatorial Pacific core V28–238. *Quaternary Research* **3**, 39–55.

Simpson, G. C. (1940). Possible causes of changes in climate and their limitations. *Proceedings of the Linnaean Society of London* **162**, 190–219.

Spitaler, R. (1921). "Das Klima der Eiszeitalters." R. Spitaler, Prague.

Street, F. A., and Grove, A. T. (1976). Environmental and climatic implications of late Quaternary lake-level fluctuations in Africa. *Nature* **251**, 385–390.

Street-Perrott, F. A. (1986). The response of lake levels to climatic change—implications for the future. *In* "Climate-Vegetation Interactions" (C. Rosenzweig and R. Dickinson, Eds.), pp. 77–80. Report OIES-2, Office for Interdisciplinary Earth Studies (OIES), University Corporation for Atmospheric Research (UCAR), Boulder, Colo.

Washington, W. M., and Parkinson, C. L. (1986). "An Introduction to Three-Dimensional Climate Modeling." University Science Books, Mill Valley, Calif., 422 pp.

Watson, R. A., and Wright, H. E., Jr. (1980). The end of the Pleistocene: A general critique of chronostratigraphic classification. *Boreas* **9**, 153–163.

Watts, W. A., and Winter, T. C. (1966). Plant macrofossils from Kirchner Marsh, Minnesota—a paleoecological study. *Geological Society of America Bulletin* **79**, 855–876.

Webb, R. S. (1990). "Late Quaternary Water-level Fluctuations in the northeastern United States." Ph.D. thesis, Brown University, 351 pp.

Webb, T. III (1985). A global paleoclimatic data base for 6000 yr B.P. DOE Technical Report 018, Carbon Dioxide Research Division, Dept. of Energy, Washington, D.C. pp. 1–155.

——. (1986). Vegetational change in eastern North America from 18,000 to 500 yr B.P. *In* "Climate-Vegetation Interactions" (C. Rosenzweig and R. Dickinson, Eds.), pp. 63–69. Report OIES-2, Office for Interdisciplinary Earth Studies, University Corporation for Atmospheric Research, Boulder, Colo.

——. (1987). The appearance and disappearance of major vegetational assemblages: Long-term vegetational dynamics in eastern North America. *Vegetatio* **69**, 177–187.

——. (1988). Eastern North America. *In* "Vegetation History" (B. Huntley and T. Webb III, Eds.), pp. 385–414. Kluwer Academic Publishers, Dordrecht.

Webb, T. III, and Bartlein, P. J. (1988). Late Quaternary climatic change in eastern North America: The role of modeling experiments and empirical studies. *Bulletin of the Buffalo Society of Natural Sciences* **33**, 3–13.

Webb, T. III, Bartlein, P. J., and Harrison, S. P. (in press). Late-Quaternary vegetation and climate change in eastern North America. *In* "Global Climates 6000 and 9000 Years Ago" (COHMAP Members, Eds.). University of Minnesota Press, Minneapolis.

Webb T. III, Bartlein, P. J., and Kutzbach, J. E. (1987). Climatic change in eastern North America during the past 18,000 years: Comparisons of pollen data with model results. *In* "The Geology of North America, Vol. K-3, North America and Adjacent Oceans during the Last Deglaciation" (W. F. Ruddiman and H. E. Wright, Jr., Eds.), Ch. 20, pp. 447–462. Geological Society of America, Boulder, Colo.

Webb, T. III, and Bryson, R. A. (1972). Late- and postglacial climate change in the northern Midwest: Quantitative estimates derived from fossil pollen spectra by multivariate statistical analysis. *Quaternary Research* **2**, 70–115.

Webb, T. III, Cushing, E. J., and Wright, H. E., Jr. (1983). Holocene changes in the vegetation of the Midwest. *In* "Late-Quaternary Environments of the United States, Vol. 2, The Holocene" (H. E. Wright, Jr., Ed.), pp. 142–165. University of Minnesota Press, Minneapolis.

Webb, T. III, and Wigley, T. M. L. (1985). What past climates can indicate about a warmer world. *In* "The Potential Climatic Effects of Increasing Carbon Dioxide" (M. C. MacCracken and F. M. Luther, Eds.), pp. 239–257. U.S. Dept. of Energy, Report DOE/ER-0237, Washington D.C.

Winkler, M. G., Swain, A. M., and Kutzbach, J. E. (1986). Middle Holocene dry period in the northern Midwestern United States: Lake levels and pollen stratigraphy. *Quaternary Research* **25**, 235–250.

Wright, H. E., Jr. (1968). History of the prairie peninsula. *In* "The Quaternary of Illinois" (R. E. Bergstrom, Ed.), pp. 78–88. College of Agriculture, University of Illinois, Special Publication 14, Urbana.

——. (1976a). Dynamical nature of Holocene vegetation. *Quaternary Research* **6**, 581–596.

——. (1976b). Environmental setting for plant domestication in the Near East. *Science* **194**, 385–389.

——. (1977). Environmental change and the origin of agriculture in the old and new world. *In* "Origins of Agriculture" (C. A. Reed, Ed.), pp. 281–318. Mouton Publishers, The Hague.

Wright, H. E., Jr., McAndrews, J. H., and van Zeist, W. (1967). Modern pollen rain in western Iran, and its relation to plant geography and Quaternary vegetational history. *Journal of Ecology* **55**, 415–443.

Wright, H. E., Jr., and Watts, W. A. (1969). Glacial and vegetational history of northeastern Minnesota. *Minnesota Geological Survey* SP-11, Minneapolis.

Wright, H. E., Jr., Winter, T. C., and Patten, H. L. (1963). Two pollen diagrams from southeastern Minnesota: Problems in the late- and post-glacial vegetational history. *Geological Society of America Bulletin* **74**, 1371–1396.

Notes on Contributors

About the Editors

Edward J. Cushing has a Ph.D. from the University of Minnesota in 1963 and has been a professor at Minnesota since 1964. He is one of the founders of the Cooperative Holocene Mapping Project and has edited a book on Quaternary paleoecology. He spent a postdoctoral year with the Danish Geological Survey and recently spent a year in Indonesia on a Fulbright Fellowship at Gadjah Mada University, teaching and developing programs in palynological research in the Javanese tropics.

Linda C. K. Shane obtained her Ph.D. in botany from Kent State University in Ohio in 1976 and has been a research associate at the Limnological Research Center at the University of Minnesota since 1978. Her current projects involve palynological studies of vegetation and climate change focusing in midwestern North America, on late-glacial climatic reversals analogous to those of the European record and on detailed Holocene records.

About the Contributors

Richard W. Battarbee has a Ph.D. from the School of Biological and Environmental Studies, The New University of Ulster, in 1973 and currently is a professor in the Department of Geography, University College London. He has been involved with multiple grants and projects on the history of water quality. Among his many recent publications is a major review prepared for the British Department of the Environment titled "Lake Acidification in the United Kingdom 1800–1986." He has been a visiting scientist at Joensuu University, Finland, and at the Limnological Research Center at the University of Minnesota.

Richard S. Clymo obtained his Ph.D. in botany from University College London in 1961 and is now professor of ecology at Queen Mary and Westfield College, University of London. His research and publications involve all aspects of peatland studies including water quality, *Sphagnum* physiology, peat formation processes, and peatland histories.

John E. Kutzbach obtained his Ph.D. in meteorology from the University of Wisconsin-Madison in 1966. He is currently director of the Center for Climatic Research and chairman of the Department of Meteorology and holds a Bascom

Professorship at the University of Wisconsin. His research has focused on the internal dynamics of global climate and global climate models, including change through geologic time. He has been a visiting scientist in England (NATO Fellowship), in Switzerland (United Nations), at the University of Bonn (Humboldt Fellowship), and at the National Center of Atmospheric Research in Boulder.

Jan Mangerud has a Ph.D. in geology from the University of Bergen in 1973 and is at present a professor of Quaternary geology there. He has published extensively in both Norwegian and English on his research into problems of glacial and biotic stratigraphy in northern Europe and the Arctic. He has been a visiting scientist at the universities of Stockholm, Minnesota, and Colorado and received the Reusch Medal from the Norwegian Geological Society.

James C. Ritchie obtained his Ph.D. in ecology from the University of Sheffield, England, in 1955 and his D.Sc. from the University of Aberdeen in 1962. Currently he is a professor in botany at Scarborough College, University of Toronto. His research interests involve both modern and late Quaternary biota in the Siberian/Alaskan region and biotic and lake-level history in North Africa and the Mediterranean.

Patty Jo Watson obtained her Ph.D. from the University of Chicago in 1959 and currently is a professor of anthropology at Washington University, St. Louis, and codirector of the Shell Mound Archaeological Project in western Kentucky. Her research has been involved with long-term archaeological field projects in Iran, Turkey, and in many North American locations focusing on problems of economic substance and environmental reconstruction. She has published nine books and currently has three in progress, one on Zuni prehistory and two on Kentucky prehistory.

Thompson Webb III obtained his Ph.D. in meteorology from the University of Wisconsin-Madison in 1971 and currently is a professor at Brown University in Rhode Island. He is one of the founding members of COHMAP, and his research has involved climatic and vegetational interpretation of the pollen record on multiple scales from a square km to global. He has been a visiting scientist at the University of Wisconsin, the University of Colorado, and the University of Cambridge. He has edited two books, both concerning vegetation and climate dynamics.

Index

Reference to a figure/figure caption is, 99*f*; to a table, 136*t*; to a mathematical expression, 157*m*.